# Recent Advances in the Assessment of Flood Risk in Urban Areas

# Recent Advances in the Assessment of Flood Risk in Urban Areas

Editor

**Tiago Miguel Ferreira**

MDPI • Basel • Beijing • Wuhan • Barcelona • Belgrade • Manchester • Tokyo • Cluj • Tianjin

*Editor*
Tiago Miguel Ferreira
University of Minho
Portugal

*Editorial Office*
MDPI
St. Alban-Anlage 66
4052 Basel, Switzerland

This is a reprint of articles from the Special Issue published online in the open access journal *Water* (ISSN 2073-4441) (available at: https://www.mdpi.com/journal/water/special_issues/Flood_Risk_Urban).

For citation purposes, cite each article independently as indicated on the article page online and as indicated below:

LastName, A.A.; LastName, B.B.; LastName, C.C. Article Title. *Journal Name* **Year**, *Article Number*, Page Range.

**ISBN 978-3-03936-830-3 (Hbk)**
**ISBN 978-3-03936-831-0 (PDF)**

Cover image courtesy of Andy Falconer.

# Contents

# About the Editor

**Tiago Miguel Ferreira** holds a PhD in Civil Engineering. His research domains include the evaluation and mitigation of urban risks—including seismic, fire, and flood—and the structural safety assessment and retrofit of ancient stone masonry buildings. He is currently a Junior Researcher at the Institute for Sustainability and Innovation in Structural Engineering (ISISE) of the University of Minho, invited Assistant Professor at the Department of Civil Engineering of the University of Coimbra, and Editor-in-Chief of *Conservar Património*, a SCOPUS and Web-of-Science indexed journal dedicated to heritage studies.

*water*

MDPI

*Editorial*

# Recent Advances in the Assessment of Flood Risk in Urban Areas

**Tiago Miguel Ferreira**

ISISE, Institute of Science and Innovation for Bio-Sustainability (IB-S), Department of Civil Engineering, University of Minho, 4800-058 Guimarães, Portugal; tmferreira@civil.uminho.pt

Received: 23 June 2020; Accepted: 25 June 2020; Published: 29 June 2020

The adverse effects of flood disasters in urban areas have been increasing in severity and extent over the past years. The amounts of losses resulting from these events are also increasing exponentially, particularly in highly urbanised urban areas, where the effects of intensive land use and climate change are particularly extreme [1]. All this despite our scientific knowledge, technical competence and computational capacity to develop highly sophisticated and accurate forecasting and simulation models being higher than ever, as is our capacity to map and analyse flood-related data.

In order to tackle this global issue, it is fundamental to keep on promoting and developing fundamental and applied research that allows the better targeting of interventions to improve resilience, reduce vulnerability and enhance recovery, as well as assisting decision-makers in delivering more effective flood risk-reduction policies. The present Special Issue of Water aims to contribute to this goal by providing a space in which to share and discuss recent studies and state-of-the-art methodologies focused on the assessment and mitigation of flood risk in urban areas. It includes nine high-quality research articles authored by eminent scholars from India, Italy, Korea, Portugal, Romania, Singapore, Spain, Taiwan, Thailand and Vietnam, who had the tremendous generosity to join me in this project. The range of topics covered by these nine studies is extraordinarily vast, reflecting the complexity of the current challenges associated with the topic.

Lee et al. [2] present an interesting discussion on the role played by drainage facilities in urban flooding. Based on the analysis of a series of past flooding events, the authors propose a resilience index to be used to diagnose the status of urban drainage systems. Such an indicator, easily calculated using real-time rainfall data, allows for the identification of intervention needs and for the evaluation of the impact of structural and non-structural interventions aimed at reducing urban inundation—thus providing a proper means of supporting more efficient decision-making processes, fostering urban resilience.

Afifi et al. [3] propose a GIS-based high-resolution flood loss and risk assessment model to analyse the impacts of flooding events in residential urban areas. For such, different flood loss, hazard and vulnerability indicators—such as water depth, elevation, distance to first responders and population density—are combined using an Analytic Hierarchy Process (AHP). After being comprehensively presented and discussed, the model is used to generate a series of hazard, vulnerability and flood risk maps for the city of Tainan in Taiwan. Among other interesting particulars, the authors provide a critical discussion on the effects (measured in terms of estimated losses) of different spatial model resolutions.

Park and Lee [4] present a study aimed at contributing to the reduction and minimisation of flood damage in the case of heavy rainfall in urban areas, by increasing urban spatial efficiency through the grading of flood risk. To this end, the authors suggest an urban flood risk assessment model able to account for different urban planning elements, namely the land use and building characteristics. The capacities and the potential of the model are discussed through its application to Changwon City in Korea. The role played by land-use planning measures to prevent flooding, as well as the possibility of classifying flood risk areas according to the land use of the districts and of evaluating intervention priorities by selecting high-risk areas, are some of the aspects discussed by the authors.

Littidej and Buasri [5] analyse the way in which the Digital Elevation Model (DEM) and Flood Risk Susceptibility (FRS) prediction models are impacted by urban growth in Muaeng District, Nakhon Ratchasima (Thailand). The scope of this study comprises three main objectives: to optimise the Cellular Automata (CA) model for predicting the expansion of built-up sites, to model a linear regression method for deriving the transition of the digital elevation model (DEM), and to apply a Geographic Weighted Regression (GWR) for analysing the risk of the stativity of flood areas in the province. Among other noteworthy outcomes, this study shows that the CA model can accurately predict the expansion of built-up areas using land-use data updated at 2-year intervals.

Mihu-Pintilie et al. [6] develop a method for flood vulnerability assessment under real (average discharge) and mathematical (calculated discharge) hydrological data based on HEC-RAS, high-density LiDAR data, and 2D hydraulic modelling. The authors compute four 2D streamflow hydraulic scenarios in order to test the flood mitigation capacity of hydro-technical constructions located downstream on the Bistriţa River, North-East Romania. With this application, the authors show that multi-scenario results obtained from 2D hydraulic modelling can be applied to obtain flood hazard parameters (such as the flood depth, flood extent, flood velocity, or water surface elevation), which can be subsequently used in recognising and responding to flood threats at the local level.

Bernardini and Quagliarini [7] address the important issue of individuals' evacuation in urban flood scenarios by providing general and unified modelling approaches to estimate evacuation speed variations depending on an individual's excitement (walking or running), floodwater depths and individuals' features (age, gender, height and their average speed on dry surfaces). Speed data from previous experiments are organized using linear regression models. Finally, the authors discuss the possible implementation of these models to simulate evacuees' motion in floodwaters (considering different confidence degree levels) and to assess the community's flood risk and the effectiveness of risk-reduction strategies.

Dhara et al. [8] discuss the suitability of using different sources of satellite imagery for capturing the extent of flood inundation in urban areas. In this proof-of-concept study, the authors show that machine-learning algorithms—namely the Support Vector Machine Regression (SVR) technique—can be efficiently used to integrate data from three different satellite sensors (Landsat, MODIS and Sentinel-2) to derive one single integrated value. Because it relies on the use of freely available data, this approach constitutes an economical and efficient alternative, mainly when field data are not readily available, which is particularly frequent in regions with difficult access or with high rates of urban growth. The city of Can Tho (Vietnam) is used to prove the concept and to illustrate the potential of the approach.

Ferreira and Santos [9] discuss the application of an integrated flood risk assessment approach, which combines flood hazard and building vulnerability indicators to identify and classify risk and to narrow intervention priorities. The Historic City Centre of Guimarães, in Portugal, is exploited by the authors to illustrate the application of the methodology. After modelling the flood hazard using the hydrologic–hydraulic method and evaluating the flood vulnerability of the buildings by resorting to a simplified vulnerability assessment method, they provide a comprehensive analysis of the outputs in both an individual and integrated manner. Finally, they also use a risk-matrix approach to aggregate these hazard and vulnerability outputs and categorise the buildings into different qualitative levels of risk.

Arosio et al. [10] analyse the direct and indirect impacts of rainfall flooding in Mexico City. To do so, they resort to a graph-based methodology in which the exposed elements are organized as nodes on a graph that is used to propagate impacts from directly affected nodes to other nodes across network links. The authors discuss how the impacts are propagated along different orders of the impact chain for increasing return periods, also comparing the risk curves between direct and indirect impacts. Finally, they highlight the extent to which the reduction in the demand for services from consumers and the loss of services from suppliers are respectively contributing to the final indirect impacts, as well as how different impact mitigation measures can be formulated from these results.

I would like to express my sincere gratitude to the authors, who have kindly shared their scientific knowledge and experience through their contributions; to the peer reviewers, who have contributed to enhancing the quality of the articles published in this Special Issue; and to the managing editors of the journal, who have supported and promoted this project. Last but not least, I would like to leave a word of appreciation to the authors who, unfortunately, saw their manuscripts declined as a result of the high-quality standards adopted in the selection and review of the received articles.

**Funding:** This research was funded by the Portuguese Foundation for Science and Technology (FCT) through the postdoctoral grant SFRH/BPD/122598/2016.

**Conflicts of Interest:** The author declares no conflict of interest.

## References

1. Miranda, F.N.; Ferreira, T.M. A simplified approach for flood vulnerability assessment of historic sites. *Nat. Hazards* **2019**, *96*, 713–730. [CrossRef]
2. Lee, E.H.; Choi, Y.H.; Kim, J.H. Real-Time Integrated Operation for Urban Streams with Centralized and Decentralized Reservoirs to Improve System Resilience. *Water* **2019**, *11*, 69. [CrossRef]
3. Afifi, Z.; Chu, H.-J.; Kuo, Y.-L.; Hsu, Y.-C.; Wong, H.-K.; Ali, M.Z. Residential Flood Loss Assessment and Risk Mapping from High-Resolution Simulation. *Water* **2019**, *11*, 751. [CrossRef]
4. Park, K.; Lee, M.-H. The Development and Application of the Urban Flood Risk Assessment Model for Reflecting upon Urban Planning Elements. *Water* **2019**, *11*, 920. [CrossRef]
5. Littidej, P.; Buasri, N. Built-Up Growth Impacts on Digital Elevation Model and Flood Risk Susceptibility Prediction in Muaeng District, Nakhon Ratchasima (Thailand). *Water* **2019**, *11*, 1496. [CrossRef]
6. Mihu-Pintilie, A.; Cîmpianu, C.I.; Stoleriu, C.C.; Pérez, M.N.; Paveluc, L.E. Using High-Density LiDAR Data and 2D Streamflow Hydraulic Modeling to Improve Urban Flood Hazard Maps: A HEC-RAS Multi-Scenario Approach. *Water* **2019**, *11*, 1832. [CrossRef]
7. Bernardini, G.; Quagliarini, E. How to Account for the Human Motion to Improve Flood Risk Assessment in Urban Areas. *Water* **2020**, *12*, 1316. [CrossRef]
8. Dhara, S.; Dang, T.; Parial, K.; Lu, X.X. Accounting for Uncertainty and Reconstruction of Flooding Patterns Based on Multi-Satellite Imagery and Support Vector Machine Technique: A Case Study of Can Tho City, Vietnam. *Water* **2020**, *12*, 1543. [CrossRef]
9. Ferreira, T.M.; Santos, P.P. An Integrated Approach for Assessing Flood Risk in Historic City Centres. *Water* **2020**, *12*, 1648. [CrossRef]
10. Arosio, M.; Martina, M.L.V.; Creaco, E.; Figueiredo, R. Indirect Impact Assessment of Pluvial Flooding in Urban Areas Using a Graph-Based Approach: The Mexico City Case Study. *Water* **2020**, *12*, 1753. [CrossRef]

*water*

MDPI

*Article*

# Real-Time Integrated Operation for Urban Streams with Centralized and Decentralized Reservoirs to Improve System Resilience

**Eui Hoon Lee** [1], **Young Hwan Choi** [2] and **Joong Hoon Kim** [3,*]

[1] School of Civil Engineering, Chungbuk National University, Cheongju 28644, Korea; hydrohydro@naver.com
[2] Research Center for Disaster Prevention Science and Technology, Korea University, Seoul 02841, Korea; yh.choi2709@gmail.com
[3] School of Civil, Environmental and Architectural Engineering, Korea University, Seoul 02841, Korea
* Correspondence: jaykim@korea.ac.kr; Tel.: +82-02-3290-3316

Received: 4 December 2018; Accepted: 26 December 2018; Published: 2 January 2019

**Abstract:** Recently, the number of extreme rainfall events has increased because of climate change. The ever-widening impervious area in urban watersheds also continuously augments runoff volume. Most measures to prevent urban inundation are structural, such as the construction, rehabilitation, and replacement of urban drainage facilities. Because structural measures require time and money, nonstructural measures are also required for the efficient prevention of urban inundation. Current operations in Korea focus on the individual operation of urban drainage facilities while neglecting the status of effluent streams. A study on urban drainage facilities that considers the status of urban streams is necessary to improve the operation of drainage facilities in urban areas. A revised resilience index is suggested to evaluate measures. For the historical rainfall event in 2010, the system resilience for current and integrated operations was 0.199 and 0.238, respectively. For the 2011 event, the system resilience for current and integrated operations was 0.064 and 0.235, respectively. The integrated operation exhibited good performance for the 2010 and 2011 events. Based on the results of this study, an operation as a nonstructural measure for the total management of urban areas is proposed. The revised resilience index could support decision-making processes for flood-management plans.

**Keywords:** integrated operation; urban stream; urban drainage facility; revised resilience index

## 1. Introduction

As the discharge in urban areas has increased because of urbanization, various measures to prevent inundation have been considered. Structural measures focus on expanding the capacity of urban drainage facilities through construction, rehabilitation, and replacement. The construction of additional drainage facilities can be classified as structural because it requires money and time. Nonstructural measures, such as the operation of urban drainage facilities (centralized and decentralized reservoirs) and flooding alerts, are required to maximize the efficient management of these structural measures. In general, the status of urban streams is not considered because urban drainage facilities in Korea are individually operated to prevent urban inundation. The management of urban drainage systems should include all components, such as urban streams, pump stations, and detention reservoirs; however, previous studies did not consider these components. The previous studies focused on the individual real-time control (RTC) and cooperative operation of urban drainage facilities in a single drainage area. The related studies of RTC in urban drainage can be classified into categories of individual operation and cooperative operation in a drainage area, as shown in Table 1.

**Table 1.** Classification of real-time control (RTC) in urban drainage systems.

| Individual Operation | Cooperative Operation in a Single Drainage Area |
|---|---|
| Beeneken et al. (2013); Cembrano et al. (2004); Galelli et al. (2012); Hsu et al. (2013); Lee et al. (2017); Raimondi and Becciu (2015); Schütze et al. (2004); Sweetapple et al. (2018); Tamoto et al. (2008); Xu et al. (2018); Zacharof et al. (2004) | Fiorelli et al. (2013); Hsu et al. (2000); Kroll (2018); Lee et al. (2016); Lund et al. (2018); Pleau et al. (2005); Vanrolleghem et al. (2005) |

Most of the studies related to individual operation have been conducted according to the unit time of operation. Individual operation has been suggested for single urban drainage facilities using various techniques [1–7]. Additionally, individual operation considering structural and other nonstructural measures has been applied to the design and forecasting field [8–11]. The cooperative operation of a single drainage area has been conducted for reducing the flooding of a city by applying the cooperation of the sewer system and the pump facility in the same drainage area. Cooperative operation with optimization, including single and multi-objective functions, has been applied to urban drainage systems [12–14]. Additional approaches for cooperative operation using the urban inundation model and the concept of resilience have been introduced [15–18].

These operations can be categorized as individual and cooperative operation of urban drainage facilities in a single drainage area. However, the integrated operation of urban drainage facilities has not yet been examined while considering the status of urban streams. Previous literature on RTC in urban drainage systems focused on the prevention of urban inundation. These approaches (i.e., individual operation and cooperative operation in a drainage area) can reduce the damage that urban inundation causes. However, it is difficult for these approaches to respond under extreme and localized torrential rainfall conditions, and they do not consider sustainable urban development in cities. Therefore, this study proposes a new operation approach that considers the overflow in urban streams and the inundation of inland drainage areas simultaneously as well as the improvement of system resilience. Therefore, it is suggested that this integrated operation could reduce inundation and improve resilience in target watersheds. The term "integrated operation" originated from a research project (Development of advanced techniques in combined inland-river systems) that was supported by the Ministry of Land, Infrastructure, and Transport of the Korean government.

In the 2000s, new approaches that considered various models and components were suggested for more sustainable and resilient cities [19–22]. These studies are based on the resilience of various factors in urban systems and have led to studies on system resilience in civil engineering, many of which have been performed in the field of water resources engineering. Ever since the concept of resilience was applied to water resources, many studies on the resilience of urban drainage systems have been conducted. A system resilience with conceptual framework and global analysis approach has been proposed for urban drainage systems [23–27]. The resilience index in these studies was calculated by comparing normal and extreme rainfall events.

System resilience is the ability of preparation for, reaction to, and recoverability against a system failure that is generated as a malfunction of system components (e.g., the pump and gate in pump stations), and the resilience index indicates system functionality. Therefore, to maintain high system functionality, the system needs to consider system resilience. For example, in urban drainage systems, if 80 out of 100 units of water can be discharged and there are 20 units of flooding volume, the system functionality is 80%. The resilience in urban drainage systems is based on the quantification of the system functionality at each time point. The resilience in urban drainage systems has been proposed in terms of, and quantified by, the rainfall amount, target area, and flooding volume [10].

One resilience index consists of the unit area, the rainfall amount considering the time of concentration, and the flooding volume [16]. Another resilience index consists of the watershed area, the total rainfall amount, and the flooding volume [10]. A drawback of these two indices is that it is difficult to highlight the differences between each value. In the previous indices, the denominator was very large compared with the flooding volume per minute in the numerator. This was because the denominator was multiplied by the rainfall amount considering the time of concentration (or the total rainfall amount) and the unit area (or the watershed area). The fundamental reason for overestimation in these two indices originates from the rainfall amount in the denominator. The duration for calculating the rainfall amount at each time point is longer than the duration of the flooding volume at each time point. If the flooding volume occurs at 60 min and the time of concentration is 30 min, the rainfall amount is calculated as the sum of 30 min to 60 min (or 1 min to 60 min). The calculation of the rainfall amount needs to be improved.

The revised resilience index can better reflect the status of urban drainage systems. It is calculated using real-time rainfall data and is easy to apply. Various conditions, such as flooding in times of no rainfall and no flooding during rainfall events, can now be considered. A revised resilience index is needed to overcome the disadvantages of the previous indices.

## 2. Materials and Methodologies

### *2.1. Integrated Operation Considering Urban Streams*

The operation of pump stations should be integrated considering the levels of monitoring nodes and urban streams within the target watershed. RTC in this study means the operation/control of urban drainage facilities considering real-time rainfall data per minute. This approach was applied to each drainage area. Monitoring nodes were selected using the method described by Lee et al. [16], and integrated operation was based on the level of urban streams. When the stream level is high, the freeboard of levees on one or both sides may not be enough to prevent flood risk. In this case, pump stations on the stream do not operate drainage pumps as early as expected; however, they reserve a certain amount of water as the operating level in the centralized reservoirs (CRs) of the pump station increases. The CR in urban drainage systems is located in the downstream of the network in a single drainage area. It receives all inflow from the network, and the inflow in the CR is discharged by drainage pumps in pump stations.

The objective of integrated operation in urban areas is to ensure rapid and safe drainage and to secure additional storage capacity. In addition, it can prevent the backwater effect that is caused by high water levels in the CR and reduce the risk of overflow in urban streams. Moreover, this operation system allows for the effective management of urban streams when the system does not have sufficient capacity. This study looks at the development of an integrated operation system considering sustainable urban management, which applies the revised system resilience index and focuses on Korean urban areas. In Korea, it is difficult to expand the width of streams or increase the height of levees because the tops of levees are used as roads. The integrated operation can balance the spare capacity between urban drainage facilities and urban streams. A schematic of the integrated operation is shown in Figure 1. The proposed approach consists of three schemes: (1) integrated pump operation, (2) the revised system resilience index, and (3) the selection of monitoring nodes.

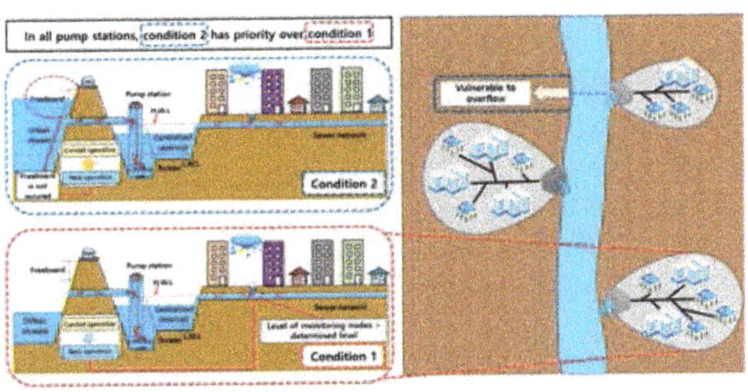


**Figure 1.** A schematic of integrated operation in urban areas. LWL, low water level; HWL, high water level.

### 2.2. Integrated Pump Operation

The operation of pump stations in Korea is determined by the level of the CR. The initial CR operating level ($O_i$) is calculated as the sum of the required depth, screen head loss ($H_s$), and freeboard for mechanical operation ($F_m$). $V_r$ is the value of the product of the initial pump capacity and pump preparation time ($T_p$). The value of $H_s$ varies from 0.1 m to 0.3 m, and the value of $F_m$ varies from 0.0 m to 0.2 m. $H_s$ is determined by the presence or absence of a CR in pump stations, and $F_m$ by the total pump capacity. Both should be determined to prevent the cavitation of drainage pumps. These three factors are added to the bottom of the CR ($B_{cr}$) to give the initial operating level of drainage pumps ($O_i$). Other operating levels, except for the initial operating level, are based on the calculation of the required depth. Early operating levels in each pump station are calculated using the method of Lee et al. [16]. Equation (1) is used to calculate the initial operating level of the drainage pumps in the CR.

$$O_i = \frac{P_i \times T_p}{4V_r A_l} + H_s + F_m + B_{cr} \quad (1)$$

where $O_i$ is the initial operating level of the drainage pumps in the CR, $P_i$ is the capacity of the initial pump, $T_p$ is the preparation time of the initial pump, $V_r$ is the required volume in the CR, $A_l$ is the average area at each elevation in the CR, $H_s$ is the screen head loss, $F_m$ is the freeboard for mechanical operation, and $B_{cr}$ is the bed elevation of the CR. Figure 2 shows the determination of operating levels for drainage pumps in CRs.

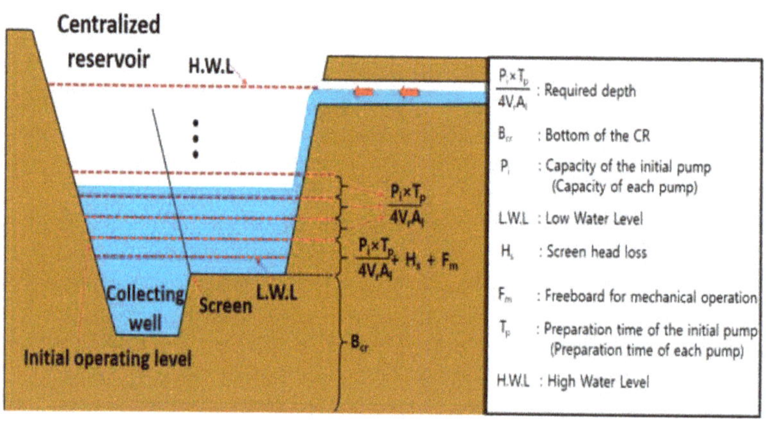


**Figure 2.** Determination of operating levels for drainage pumps in the centralized reservoir (CR).

As the local government in Korea has already determined the operating order for all pump stations, the new operation in the CR has the same operating order as the current one. Pump operation should be determined by operating level and order, considering the cavitation of drainage pumps. When the level of a monitoring node reaches a determined level for a new operation, drainage pumps in the CR are operated early. The operation of the decentralized reservoir (DR) is applied to the integrated operation if there is a DR in the drainage area. The DR's operation is also based on the level of the monitoring nodes. Pumps in the DR are operated until the level of the monitoring node reaches the operating level of the CR. Lee et al. [16] suggested the cooperative operation of the CR and DR, which means the combined operation of the CR and the DR. In this study, we propose an integrated operation for urban streams with all drainage facilities, including the CR and the DR.

### 2.3. Revised Resilience Index

Resilience is the ability to recover from a failure because of internal and external shocks to a community system. It involves the ability to recover from flooding or a malfunction of drainage facilities in urban drainage systems. Recently, the concept of resilience has been used in the evaluation of various systems, including future strategies by governments, academia, and enterprises. Todini [28] applied the resilience index to the design of a looped water distribution network using a heuristic approach. A framework to quantitatively assess the improvement in a community's resilience to seismic hazards was developed by Bruneau et al. [29]. Godshalk [30] recommended constructing resilient cities to withstand the threats of natural hazards and terrorism.

Structural and functional resilience indices in urban drainage systems were suggested in [25,26]. These consist of the total flood volume, the total inflow into the system, the mean duration of nodal flooding (computed for all flooded nodes in the system), and the maximum nodal flood duration or total elapsed time (simulation time). The resilience index in the two studies is the computed design functional resilience index for the design of urban drainage systems. It is not based on real rainfall but on rainfall scenarios. In addition, its application is very complicated because it requires the estimation of the areal reducing factor. Lee et al. [10] introduced the resilience index for urban drainage systems, which is based on the flooding volume per minute, the area of the target watershed, and the total rainfall amount. In this index, the flooding volume has a low impact if the area of the target watershed and the total rainfall amount are large. All values of the resilience index in the previous study are close to 1; therefore, it is difficult to see a difference between current and new measures. In this study, a revised resilience index is introduced to overcome this difficulty.

The main difference between the previous and revised resilience indices lies in the calculation of the denominator in the performance evaluation function. In the previous resilience index, the total rainfall amount is applied to the denominator, while in the revised resilience index, the rainfall amount at each minute is applied to the denominator. The value of the denominator in the previous resilience index is larger than that in the revised resilience index. In addition, these values are classified into various cases because the denominator cannot be calculated when the rainfall amount is zero in the revised resilience index. The revised resilience is calculated as 1 when the rainfall amount and flooding volume are zero. The revised resilience is calculated as 0 when the rainfall amount is zero and the flooding volume is not zero. The rainfall amounts in both resilience indices are different, even though the performance evaluation function is calculated every minute in both. The total amount of rainfall in the watershed is used in the previous resilience index, while the current rainfall amount in the watershed is used in the revised one. The resilience index comprises the performance evaluation function, which is shown in Equation (2):

$$u(T)_t = \max\left(0, 1 - \frac{F_t}{R_t \times A}\right) \tag{2}$$

where $u(T)_t$ is the performance at time $t$, $F_t$ (in m$^3$) is the flooding volume at time $t$, $R_t$ is the rainfall amount (in m) at time $t$, and $A$ is the area of the target watershed (in m$^2$). If the value of the performance evaluation is 1, then there is no flooding. Table 2 categorizes the value of utility performance according to the numerator and denominator of Equation (2). In the revised index, the duration of failure is calculated in the equation for resilience because the performance evaluation function is estimated at each minute and the equation for resilience is based on the value of the performance evaluation function.

**Table 2.** The value of the performance evaluation according to the numerator and denominator.

| Value of Numerator ($F_t$) | Value of Denominator ($R_t \times A$) | Status of System | Value of Performance Evaluation Function |
|---|---|---|---|
| 0 | 0 | No failure | 1 |
| 0 | Over 0 | No failure | 1 |
| Over 0 | 0 | Failure | 0 |
| Large | Small | Failure | 0 |
| Small | Large | Partial Failure | $1 - \frac{F_t}{R_t \times A}$ |

The sum of the numerators at each time point gives the total flooding volume in the target watershed, and the sum of the denominators at each time point gives the total rainfall amount. The resilience in the target watershed is calculated by the value of the performance every minute. The equation for resilience is shown in Equation (3).

$$R_s = \frac{1}{T}\int_0^T u(T)_t dt \tag{3}$$

where $R_s$ is the resilience in the target watershed, and $T$ is the entire duration of the rainfall event. The revised resilience index can be calculated for and applied to all watersheds.

### 2.4. Selection of Monitoring Nodes

Monitoring nodes are points where inundation first occurs within an urban drainage system [10,16]. Three synthetic rainfall durations were selected on the basis of the time of concentration (tc) (e.g., tc, 2tc, and 3tc).

Time of concentration was used as a variable to allow for a sufficient duration of rainfall to be simulated considering the total amount of rainfall in the drainage system. Additional durations, up to three times as long as the time of concentration, were selected to simulate longer-duration rainfall events. For example, if the time of concentration is 30 min, the three durations are 30, 60, and 90 min. Initially, 1 mm of synthetic rainfall data were produced. This was then increased in 1-mm increments until the first flooding event occurred. The process for finding the first flooding nodes is shown in Figure 3.

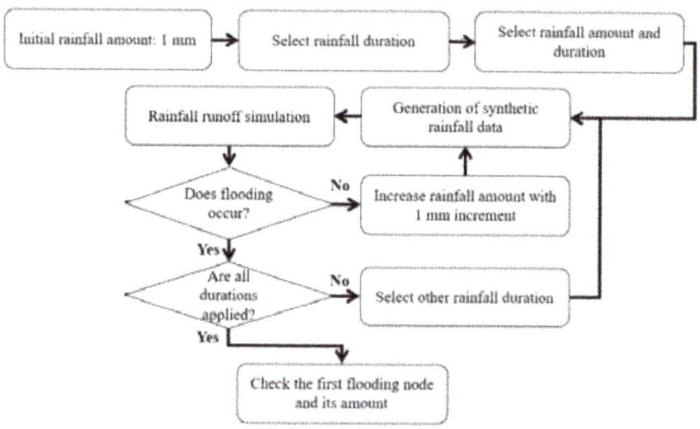


**Figure 3.** The process for finding the first flooding nodes.

A report detailing the design of discharge in streams in Korea is available to the public [31]. Most streams in Korea have several monitoring nodes under bridges. Monitoring nodes for integrated operation can be selected from among these. Various discharges for each frequency can be applied to the selected simulation model, and the freeboards for each monitoring node can be checked. The monitoring node with the smallest value of freeboard is then selected for use in the integrated operation. The schematic of the monitoring nodes in streams for integrated operation is shown in Figure 4.

In Figure 4, among the four candidates, candidate B can be selected as the monitoring node because the freeboard at candidate B is smaller than the required freeboard. The candidate with the largest difference is selected if the freeboard at several candidates is smaller than the required freeboard. In Korea, monitoring candidates are generally located on bridges, and the required freeboard at each station of streams is determined by the design frequency. The required freeboard of the stream is high if the design frequency of the stream is high. For example, the required freeboard of the stream is 0.6 m for a 100-year frequency, and the required freeboard is 0.8 m for a 200-year frequency. The design frequency of the downstream is larger than that of the upstream if the design frequencies of the upstream and downstream are different.

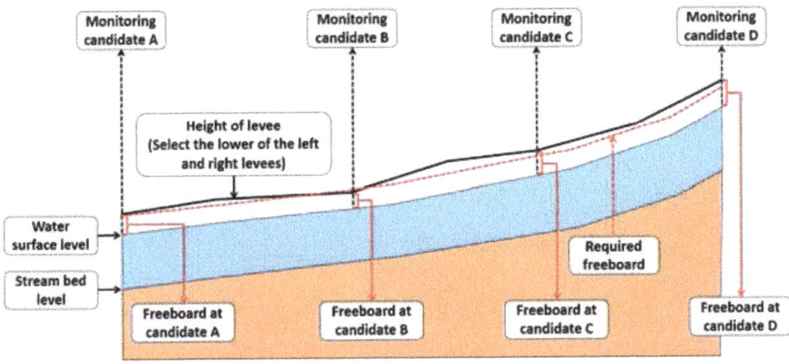


**Figure 4.** A schematic of the monitoring nodes in streams for integrated operation.

## 2.5. Model Formulation

This study consisted of five parts, as shown in Figure 5. First, synthetic rainfall data were generated for the selection of monitoring nodes. Second, monitoring nodes in each drainage area were selected to operate drainage pumps in the CR. Third, a monitoring node in an urban stream was selected for integrated operation in each CR. Fourth, the integrated operation of pump stations was conducted using the level of a monitoring node in an urban stream. Finally, a revised resilience index was suggested and applied to urban drainage systems in a target watershed.

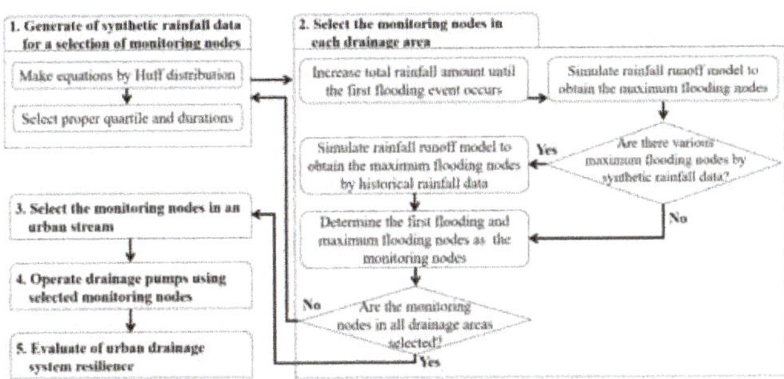


**Figure 5.** The flowchart for this study.

A storm water management model (SWMM) was used for the rainfall runoff simulation of the target watershed [32]. Evaporation and losses can be simulated in the SWMM. In Korea, the evaporation data are generally provided per month and not per minute. Monthly averages in evaporation can be applied to the SWMM, which is only used during dry periods. For infiltration among losses, Horton's infiltration equation was used. Each parameter was changed during calibration. Other losses were considered using the percentage of impervious area at each sub-catchment.

## 2.6. Study Area

The Dorim stream was selected as the target watershed, and is shown in Figure 6. It has a total area of 41.93 $km^2$. Furthermore, there are 11 pump stations (Mullae, Dorim2, Daerim3, Daerim2, Guro1, Guro2, Guro3, Guro4, Sinlim1, Sinlim2, and Sinlim5), two detention reservoirs (Daerim and Gwanak), and two branches (the Daebang and Bongchun streams) along the stream.

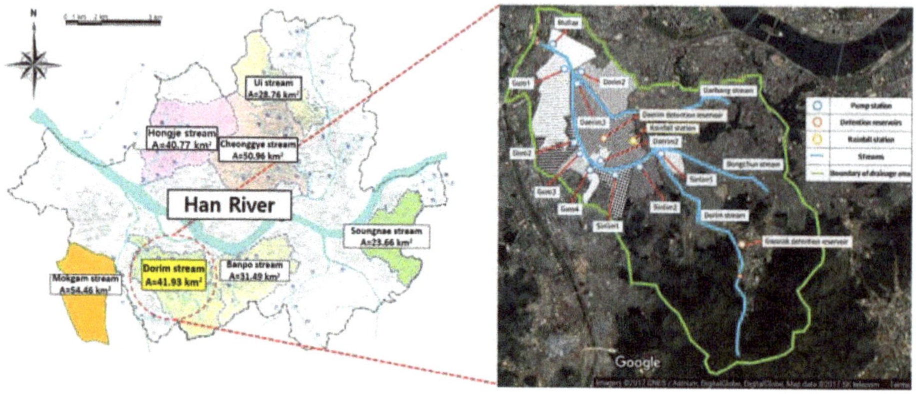


**Figure 6.** Description of the target area (Imagery © 2017 Centre National d'Etudes Spatiales/Astrium, DigitalGlobe, DigitalGlobe, Map data © SK telecom).

The characteristics of Dorim stream consist of river length, levee length, mean width of basin, shape factor, and slope. Among them, the mean width of basin is the value of area (A) divided by the river length (L) and the shape factor is the value of area (A) divided by the square of river length (L). Table 3 shows the characteristics of the Dorim stream [31].

**Table 3.** The characteristics of the Dorim stream.

| Stream | River Length (km) | Levee Length (km) | Mean WIDTH of Basin (km, A/L) | Shape Factor ($A/L^2$) | Slope |
|---|---|---|---|---|---|
| Dorim | 14.2 | 22.2 | 2.95 | 0.21 | 1/1.163~1/142 |

A rainfall station is located in the center of the target watershed. Table 4 provides information on the pump stations in the target watershed, including CR capacity, drainage pump capacity, drainage area, high water level (HWL), and low water level (LWL) [33–44].

**Table 4.** Information on pump stations in the target watershed.

| Pump Stations | Capacity of CR in Pump Stations ($m^3$) | Capacity of Drainage Pumps ($m^3$/min) | Drainage Area (ha) | Boundary Water Level |
|---|---|---|---|---|
| Guro1 | 29,100 | 1750 (245 $m^3$/min × 3, 140 $m^3$/min × 2, 270 $m^3$/min × 2, 55 $m^3$/min × 1) | 136 | HWL: 8.5 m<br>LWL: 6.0 m |
| Guro2 | 3500 | 1315 (180 $m^3$/min × 3, 360 $m^3$/min × 2, 55 $m^3$/min × 1) | 47 | HWL: 8.5 m<br>LWL: 6.5 m |
| Guro3 | 1400 | 605 (175 $m^3$/min × 3, 40 $m^3$/min × 2) | 45 | HWL: 9.5 m<br>LWL: 6.3 m |
| Guro4 | 29,100 | 480 (140 $m^3$/min × 3, 60 $m^3$/min × 1) | 27 | HWL: 9.5 m<br>LWL: 5.0 m |
| Sinlim1 | 8000 | 1000 (200 $m^3$/min × 5) | 56 | HWL: 13.7 m<br>LWL: 7.7 m |
| Sinlim2 | 5300 | 800 (64 $m^3$/min × 1, 184 $m^3$/min × 4) | 46 | HWL: 13.6 m<br>LWL: 10.5 m |
| Sinlim5 | 1000 | 411 (137 $m^3$/min × 3) | 27 | HWL: 18.2 m<br>LWL: 15.6 m |
| Mullae | 3400 | 1435 (329 $m^3$/min × 4, 119 $m^3$/min × 1) | 82 | HWL: 7.0 m<br>LWL: 3.7 m |
| Dorim2 | 31,000 | 1745 (236 $m^3$/min × 4, 267 $m^3$/min × 3) | 150 | HWL: 9.5 m<br>LWL: 7.3 m |
| Daerim2 | 1000 | 336 (336 $m^3$/min × 3) | 19 | HWL: 11.4 m<br>LWL: 8.4 m |
| Daerim3 | 36,200 | 3411 (223 $m^3$/min × 7, 150 $m^3$/min × 1, 250 $m^3$/min × 2, 600 $m^3$/min × 2) | 249 | HWL: 9.0 m<br>LWL: 6.8 m |

The total drainage area of the 11 pump stations is 884 ha (8.84 $km^2$), which is approximately 21% of the total drainage area of the Dorim stream. The Daerim detention reservoir was constructed in 2009 and the Gwanak detention reservoir in 2016. They are described in Table 5 [45,46].

**Table 5.** Description of the detention reservoirs in the target watershed.

| Detention Reservoirs | Capacity of Decentralized Reservoirs ($m^3$) | Capacity of Drainage Pumps ($m^3$/min) | Inlet Type (B × H) | Effluent Stream |
|---|---|---|---|---|
| Daerim | 2447 | 18 (9 $m^3$/min × 2) | Weir (2.0 m × 0.4 m) | Sewer network in the Daerim3 pump station |
| Gwanak | 65,000 | 12 (4 $m^3$/min × 3) | Flap gate (1.4 m × 1.4 m) | Dorim stream |

Inundations occurred at the Dorim stream in 2010 and 2011. The total annual precipitation in 2010 was 2.075 mm, and the total amount of rainfall was 253 mm when the flooding occurred on 21 September 2010. The total annual precipitation in 2011 was 2.014 mm, and the total amount of rainfall was 378 mm when the flooding occurred on 27 July 2011 [47]. These historical rainfall events have a frequency of approximately 100 years, which is greater than the design frequency of the urban drainage facilities [47,48].

The new model of the sewer network in the target watershed is based on the geographic information system data for urban drainage areas and streams supplied by the Seoul Metropolitan Government. Figure 7 shows the digital elevation model, hill shade, slope, and aspect of flow direction in the target watershed [45].

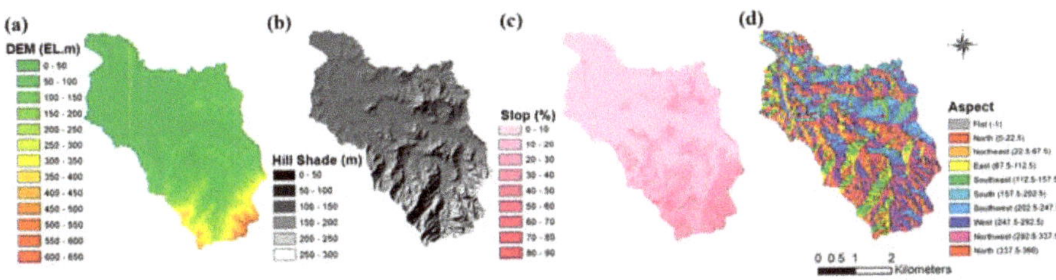


**Figure 7.** Information on the target watershed: (**a**) Digital elevation model (DEM), (**b**) Hill shade, (**c**) Slope, and (**d**) Aspect of flow direction.

As shown in Figure 8, the sewer network in the target watershed consists of 4137 sub-catchments, 4544 nodes, and 4710 links.

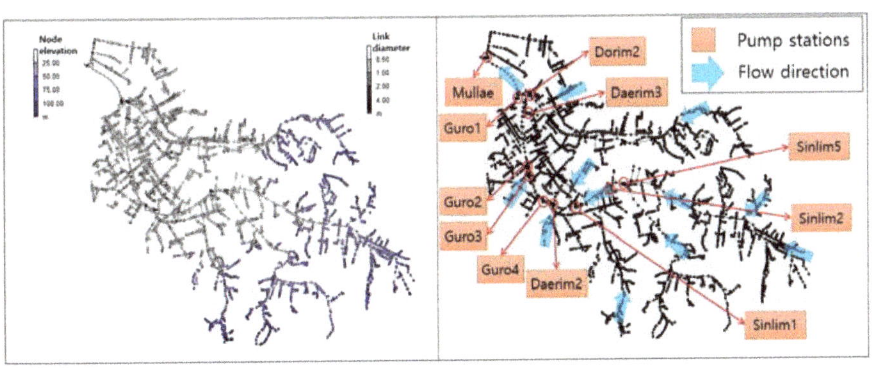


**Figure 8.** The sewer network in the target watershed.

### *2.7. Generation of Synthetic Rainfall Data*

For each drainage area and urban stream, synthetic rainfall data were generated using the Huff distribution [49], and monitoring nodes were selected by means of a rainfall runoff simulation. The synthetic rainfall data were used for the selection of monitoring nodes, and the historical rainfall data were selected to verify the effect of the integrated operation. The Huff distribution was used because its third quartile is appropriate for the design and operation of drainage facilities in Korea [50]. The design of Korean urban drainage facilities is based on this distribution. Equation (4) shows the regression equation for the third quartile of the Huff distribution in Seoul [51].

$$y = 37.835x^6 - 106.21x^5 + 105.18x^4 - 44.549x^3 + 9.1084x^2 - 0.3603x + 0.0005 \quad (4)$$

where $y$ is the ratio of cumulative rainfall, and $x$ is the ratio of the total rainfall duration. The process of generating synthetic rainfall data from the Huff distribution consists of three steps. First, a cumulative distribution is generated using the regression equation of the Huff distribution. Second, the separated ratio is produced by the cumulative distribution. Third, the rainfall amount and duration are applied to the separated ratio [51]. This process is illustrated in Figure 9.

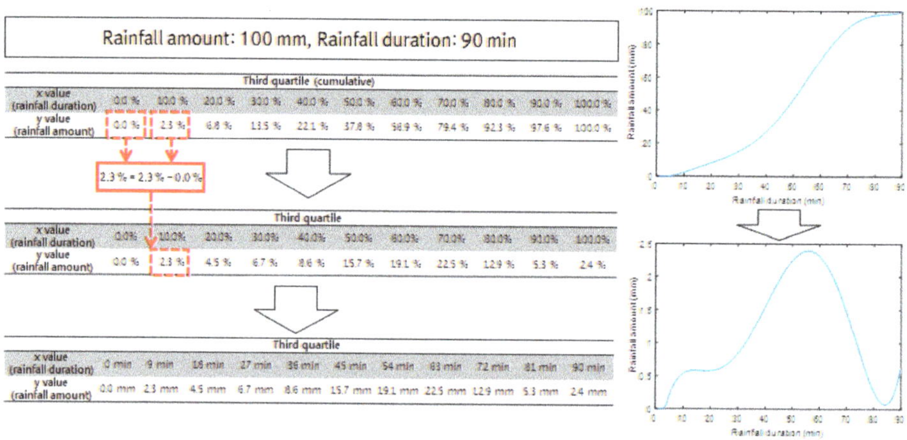


**Figure 9.** Generation of synthetic rainfall data by the Huff distribution.

Moreover, the sewer network in the target watershed was manually calibrated using the 2010 and 2011 rainfall events. After the calibration of each sub-watershed in each CR, the total runoff was calibrated downstream of each drainage area. The peak discharge, peak time, total runoff volume, and minimization of the root mean square error were focused on the calibration of the model. The percentage of impervious area, width, and percentage of slope in each sub-catchment were used to calibrate the model. In addition, the maximum infiltration rate, minimum infiltration rate, and decay constant in Horton's infiltration equation were used to calibrate the model.

## 3. Application and Results

### 3.1. Selecting Monitoring Nodes in Each Drainage Area and Stream

The selection of monitoring nodes is required for the integrated operation of drainage facilities, such as a CR and a DR. This is because a CR's operation, or a cooperative operation between a CR and a DR, is based on the level of monitoring nodes. In this study, monitoring nodes were selected using two methods, based on the first flooding node and the maximum flooding node.

The first flooding node generally occurs between branch conduits, rather than between main conduits. This makes it difficult to use for the integrated operation of drainage facilities. A section is categorized as the main conduit based on the product of the runoff coefficient (C) and drainage area (A). If this is greater than 0.12 km$^2$ (CA $\geq$ 0.12 km$^2$), then it is the main conduit. If it is smaller than 0.12 km$^2$ (CA < 0.12 km$^2$), then it is a branch [16]. The first flooding node is selected using the results of rainfall runoff simulations, with synthetic rainfall events generated by the Huff distribution.

The amount of synthetic rainfall is increased from 1 mm in 1 mm increments, and this is applied to the runoff model until the first flooding event occurs. The first flooding nodes in each drainage area are shown in Table 6.

**Table 6.** The results of the first flooding nodes for the operation of drainage facilities in each drainage area.

| Duration (min) | Guro1 (Rainfall Amount) | Guro2 (Rainfall Amount) | Guro3 (Rainfall Amount) | Guro4 (Rainfall Amount) | Sinlim1 (Rainfall Amount) | Sinlim2 (Rainfall Amount) | Sinlim5 (Rainfall Amount) | Mullae (Rainfall Amount) | Dorim2 (Rainfall Amount) | Daerim2 (Rainfall Amount) | Daerim3 (Rainfall Amount) |
|---|---|---|---|---|---|---|---|---|---|---|---|
| 30 | GR1_043 (61 mm) | GR2_082 (72 mm) | GR3_430 (65 mm) | GR4_401 (55 mm) | SL1_810 (28 mm) | SL2_420 (25 mm) | SL5_770 (27 mm) | MR_216 (38 mm) | DO_316 (49 mm) | DR2_331 (54 mm) | DR3_560 (79 mm) |
| 60 | GR1_043 (83 mm) | GR2_082 (94 mm) | GR3_430 (87 mm) | GR4_401 (76 mm) | SL1_810 (48 mm) | SL2_420 (47 mm) | SL5_770 (47 mm) | MR_216 (58 mm) | DO_316 (71 mm) | DR2_331 (76 mm) | DR3_560 (104 mm) |
| 90 | GR1_056 (128 mm) | GR2_082 (137 mm) | GR3_430 (133 mm) | GR4_401 (123 mm) | SL1_810 (96 mm) | SL2_420 (93 mm) | SL5_770 (97 mm) | MR_216 (109 mm) | DO_335 (117 mm) | DR2_331 (118 mm) | DR3_575 (154 mm) |

The rainfall amount required to cause the first flood of the drainage system in the Daerim3 pump station is higher than that in the others. This means that the drainage system in the Daerim3 pump station is relatively strong against initial flooding. Conversely, the drainage systems in the Sinlim1, Sinlim2, and Sinlim5 pump stations are relatively weak against initial flooding. Some drainage areas have different first flooding nodes for different rainfall durations. The drainage areas in Guro1, Dorim2, and Daerim3 show different first flooding nodes at 90 min because some conduits in these areas have reverse gradients, causing different initial flooding patterns. However, the Guro2, Guro3, Guro4, Guro4, Sinlim1, Sinlim2, Sinlim5, Mullae, and Daerim2 pump stations all demonstrated the same first flooding node for all durations. When a drainage area had different first flooding nodes at 90 min, the node that appeared in the largest number of results was selected.

The maximum and first flooding nodes were selected as the monitoring nodes for the integrated operation in each drainage area. To select the maximum flooding nodes, historical rainfall events were used, rather than synthetic ones, as simulations using synthetic rainfall events produce various maximum flooding nodes, making the selection of monitoring nodes difficult. Rainfall data from 23 September 2010 and 27 July 2011, when historical flooding occurred in the target watershed, were used to identify the maximum flooding nodes in each drainage area. The maximum flooding nodes in each drainage area are shown in Table 7.

**Table 7.** The results of maximum flooding nodes for the operation of drainage facilities in each drainage area.

| Rainfall Events | Guro1 | Guro2 | Guro3 | Guro4 | Sinlim1 | Sinlim2 | Sinlim5 | Mullae | Dorim2 | Daerim2 | Daerim3 |
|---|---|---|---|---|---|---|---|---|---|---|---|
| 2010 | GR1_123 (366 $m^3$) | GR2_082 (97 $m^3$) | GR3_120 (5 $m^3$) | GR4_471 (2 $m^3$) | SL1_624 (235,727 $m^3$) | SL2_420 (242,584 $m^3$) | SL5_021 (56,009 $m^3$) | MR_201 (79,380 $m^3$) | DO_316 (35,271 $m^3$) | DR2_326 (32 $m^3$) | DR3_550 (1597 $m^3$) |
| 2011 | GR1_123 (703 $m^3$) | GR2_082 (39 $m^3$) | GR3_120 (2 $m^3$) | GR4_471 (1 $m^3$) | SL1_624 (190,725 $m^3$) | SL2_420 (370,703 $m^3$) | SL5_021 (87,138 $m^3$) | MR_201 (64,473 $m^3$) | DO_316 (74,934 $m^3$) | DR2_326 (135 $m^3$) | DR3_550 (143 $m^3$) |

Several maximum flooding nodes demonstrated a greater flooding volume than those in other drainage areas, namely Sinlim1, Sinlim2, Sinlim5, Mullae, and Dorim2. This is because of the capacity shortage of conduits and backwater effects produced by the level of the CR. A single node is selected as a monitoring node if the first flooding node is the same as the maximum flooding node. Two nodes are selected as monitoring nodes if the two are different. In the real-time integrated operation, Guro1, Guro3, Guro4, Sinlim1, Sinlim5, Mullae, Daerim2, and Daerim3 each have two monitoring nodes, while Guro2, Sinlim2, and Dorim2 have one. If the depths of the two monitoring nodes differ, they are converted into a dimensionless parameter. The depth of the monitoring node is converted to 1.0D if it is 1.5 m, and the level is converted to 0.6D if it is 0.9 m.

In Korea, monitoring nodes in urban streams are constructed under bridges. There are six bridges across the Dorim stream: the Dorim, Guro1, Sindaebang, Gwanakdorim, Sinlim3, and Seoul National University Bridges. All streams in Korea have a designed freeboard, and this is 0.6 m for the Dorim stream. The monitoring candidates in the Dorim stream are shown in Figure 10.

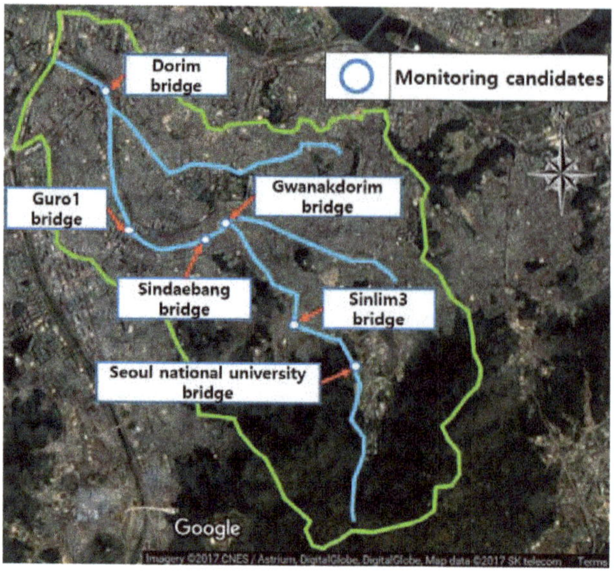


**Figure 10.** The monitoring candidates for the Dorim stream (Imagery © 2017 Centre National d'Etudes Spatiales/Astrium, DigitalGlobe, DigitalGlobe, Map data © SK telecom).

Rainfall events with various frequencies (30, 50, 80, 100, and 200 years) and with a duration based on the time of concentration in the Dorim stream (360 min) were applied to select the monitoring nodes. The water level in the Dorim stream, height of the bank, and freeboard for events with 30-, 50-, 80-, 100-, and 200-year frequencies are shown in Table 8.

**Table 8.** Information on the Dorim stream.

| Monitoring Candidates | | Water Level in Dorim Stream (EL. m) | Height of Bank (EL. m) | | Freeboard (m) | |
|---|---|---|---|---|---|---|
| | | | Left | Right | Left | Right |
| Dorim Bridge | 30-year | 12.7 | 14.36 | 14.56 | 1.66 | 1.86 |
| | 50-year | 13.22 | 14.36 | 14.56 | 1.14 | 1.34 |
| | 80-year | 13.71 | 14.36 | 14.56 | 0.65 | 0.85 |
| | 100-year | 13.92 | 14.36 | 14.56 | 0.44 | 0.64 |
| | 200-year | 14.63 | 14.36 | 14.56 | −0.27 | −0.07 |
| Guro1 Bridge | 30-year | 14.61 | 17.57 | 17.57 | 2.96 | 2.96 |
| | 50-year | 15.34 | 17.57 | 17.57 | 2.23 | 2.23 |
| | 80-year | 16.02 | 17.57 | 17.57 | 1.55 | 1.55 |
| | 100-year | 16.27 | 17.57 | 17.57 | 1.3 | 1.3 |
| | 200-year | 16.83 | 17.57 | 17.57 | 0.74 | 0.74 |
| Sindaebang Bridge | 30-year | 16.32 | 18.17 | 17.14 | 1.85 | 0.82 |
| | 50-year | 16.85 | 18.17 | 17.14 | 1.32 | 0.29 |
| | 80-year | 17.44 | 18.17 | 17.14 | 0.73 | −0.3 |
| | 100-year | 17.67 | 18.17 | 17.14 | 0.5 | −0.53 |
| | 200-year | 18.2 | 18.17 | 17.14 | −0.03 | −1.06 |
| Gwanakdorim Bridge | 30-year | 17.8 | 19.5 | 19.5 | 1.7 | 1.7 |
| | 50-year | 18.16 | 19.5 | 19.5 | 1.34 | 1.34 |
| | 80-year | 18.56 | 19.5 | 19.5 | 0.94 | 0.94 |
| | 100-year | 18.74 | 19.5 | 19.5 | 0.76 | 0.76 |
| | 200-year | 19.16 | 19.5 | 19.5 | 0.34 | 0.34 |

Table 8. *Cont.*

| Monitoring Candidates | | Water Level in Dorim Stream (EL. m) | Height of Bank (EL. m) | | Freeboard (m) | |
|---|---|---|---|---|---|---|
| | | | Left | Right | Left | Right |
| Sinlim3 Bridge | 30-year | 36.26 | 37.45 | 37.05 | 1.19 | 0.79 |
| | 50-year | 36.4 | 37.45 | 37.05 | 1.05 | 0.65 |
| | 80-year | 36.52 | 37.45 | 37.05 | 0.93 | 0.53 |
| | 100-year | 36.58 | 37.45 | 37.05 | 0.87 | 0.47 |
| | 200-year | 36.73 | 37.45 | 37.05 | 0.72 | 0.32 |
| Seoul National University Bridge | 30-year | 68.99 | 74.47 | 70.17 | 5.48 | 1.18 |
| | 50-year | 69.03 | 74.47 | 70.17 | 5.44 | 1.14 |
| | 80-year | 69.19 | 74.47 | 70.17 | 5.28 | 0.98 |
| | 100-year | 69.23 | 74.47 | 70.17 | 5.24 | 0.94 |
| | 200-year | 69.36 | 74.47 | 70.17 | 5.11 | 0.81 |

For the 30-year frequency, all monitoring candidates satisfy the designed freeboard of the Dorim stream. For the 80- and 100-year frequencies, the right bank at Sindaebang Bridge lacks a freeboard. Overflow occurs here with a 100-year frequency. The other monitoring candidates satisfy the designed freeboard of the Dorim stream. However, the banks of various monitoring candidates, such as the left bank at the Dorim Bridge, both banks at the Sindaebang Bridge, and the right bank at the Sinlim3 Bridge, also lack a freeboard. Overflow occurs at the right bank at the Sindaebang Bridge, and the overflow from a 100-year frequency event is the same as that for an 80-year frequency event. Both banks at the Dorim, Sindaebang, and Gwanakdorim Bridges, and the right bank at the Sinlim3 Bridge, also lack a freeboard. Overflow occurs at the Dorim and Sindaebang Bridges. The results in Table 8 shows that overflows at Sindaebang Bridge occur with 80-, 100-, and 200-year frequencies. This section of the Dorim stream is most vulnerable to overflow.

### 3.2. Results of the Rainfall Runoff Simulation

Historical rainfall events in 2010 and 2011, when flooding occurred in the Dorim stream, were selected for the rainfall runoff simulation, in which the current and integrated approaches for operating drainage facilities, including CR and DR operations, were applied to the target watershed. The integrated operation, which includes the use of the CR, was applied to each pump station in the Dorim stream. The preparation time was between 5 and 30 min. The calculation was used to determine the initial operating level in the CR. This was applied to the Daerim3 pump station as follows. The product of the initial pump discharge (233 $m^3$/min) and the preparation time (30 min) were divided by 4. The required volume in the CR was 1711 $m^3$ and the average area at each elevation in the CR was 11,400 $m^2$. The required depth was calculated by dividing the required volume in the CR by the average area at each elevation [16]. Table 9 shows the operating levels of the drainage pumps in each drainage area of the Dorim stream.

These data are shown for both the integrated and current operations. The results, shown in Figure 11a, indicate that the integrated operation produces a lower flooding volume than the current operation (which produces 2,905,874 $m^3$).

**Table 9.** The current and new operations of the CR for each drainage area of the Dorim stream.

| Pump Station | Operation | Operating Level (m) | | | | | | | | | | | | |
|---|---|---|---|---|---|---|---|---|---|---|---|---|---|---|
| Guro1 | Elevation (m) | 6.75 | 6.85 | 6.95 | 7.05 | 7.35 | 7.55 | 7.65 | 7.75 | 7.85 | 8.50 | - | - | - |
| | Current | - | - | 3.93 | 8.02 | 12.10 | 14.43 | 16.77 | 21.27 | 25.77 | 26.68 | - | - | - |
| | New | - | 3.93 | 8.02 | 12.1 | 14.43 | 16.77 | 21.27 | 25.77 | 26.68 | 26.68 | - | - | - |
| Guro2 | Elevation (m) | 6.7 | 7.0 | 7.1 | 7.3 | 7.8 | 7.9 | 8 | 8.2 | - | - | - | - | - |
| | Current | - | - | 12.00 | 12.00 | 12.92 | 15.92 | 18.92 | 21.92 | - | - | - | - | - |
| | New | - | 12.00 | 12.92 | 15.92 | 18.92 | 21.92 | 21.92 | 21.92 | - | - | - | - | - |
| Guro3 | Elevation (m) | 6.4 | 6.5 | 6.6 | 6.7 | 6.9 | 7.2 | 8.0 | - | - | - | - | - | - |
| | Current | - | - | 0.67 | 1.33 | 4.25 | 7.17 | 10.08 | - | - | - | - | - | - |
| | New | - | 0.67 | 1.33 | 4.25 | 7.17 | 10.08 | 10.08 | - | - | - | - | - | - |
| Guro4 | Elevation (m) | 5.3 | 5.5 | 5.9 | 6.1 | 6.4 | 7.5 | - | - | - | - | - | - | - |
| | Current | - | - | 1.00 | 3.33 | 5.67 | 8.00 | - | - | - | - | - | - | - |
| | New | - | 1.00 | 3.33 | 5.67 | 8.00 | 8.00 | - | - | - | - | - | - | - |
| Sinlim1 | Elevation (m) | 7.7 | 7.8 | 8.0 | 9.0 | 9.5 | 10.0 | 12.0 | - | - | - | - | - | - |
| | Current | - | - | - | 3.33 | 6.67 | 10.00 | 16.67 | - | - | - | - | - | - |
| | New | - | 3.33 | 6.67 | 10.00 | 16.67 | 16.67 | 16.67 | - | - | - | - | - | - |
| Sinlim2 | Elevation (m) | 10.8 | 10.9 | 11.1 | 11.3 | 11.5 | 11.7 | 13.0 | - | - | - | - | - | - |
| | Current | - | - | 1.08 | 4.15 | 7.22 | 10.28 | 13.25 | - | - | - | - | - | - |
| | New | - | 1.08 | 4.15 | 7.22 | 10.28 | 13.25 | - | - | - | - | - | - | - |
| Sinlim5 | Elevation (m) | 15.8 | 16.0 | 16.5 | 16.6 | 16.8 | 17.0 | 18.0 | - | - | - | - | - | - |
| | Current | - | - | - | - | 2.50 | 4.67 | 6.83 | - | - | - | - | - | - |
| | New | - | 2.50 | 4.67 | 6.83 | 6.83 | 6.83 | 6.83 | - | - | - | - | - | - |
| Mullae | Elevation (m) | 4.5 | 4.9 | 5.0 | 5.1 | 5.2 | 5.3 | 7.0 | - | - | - | - | - | - |
| | Current | - | - | 5.48 | 10.97 | 16.45 | 21.93 | 23.92 | - | - | - | - | - | - |
| | New | - | 5.48 | 10.97 | 16.45 | 21.93 | 23.92 | 23.92 | - | - | - | - | - | - |
| Dorim2 | Elevation (m) | 7.5 | 7.6 | 7.8 | 7.9 | 8.0 | 8.1 | 8.1 | 8.3 | 8.4 | 9.0 | - | - | - |
| | Current | - | - | - | 3.93 | 7.87 | 11.80 | 15.73 | 20.18 | 24.63 | 29.08 | - | - | - |
| | New | - | 3.93 | 7.87 | 11.80 | 15.73 | 20.18 | 24.63 | 29.08 | 29.08 | 29.08 | - | - | - |
| Daerim2 | Elevation (m) | 8.5 | 9.5 | 10.5 | 10.9 | 11.4 | - | - | - | - | - | - | - | - |
| | Current | - | - | 1.87 | 3.73 | 5.60 | - | - | - | - | - | - | - | - |
| | New | - | 1.87 | 3.73 | 5.60 | 5.60 | - | - | - | - | - | - | - | - |
| Daerim3 | Elevation (m) | 6.5 | 6.8 | 7.2 | 7.3 | 7.5 | 7.6 | 7.7 | 7.8 | 7.9 | 8.0 | 8.1 | 8.3 | 9.0 |
| | Current | - | - | - | 3.88 | 8.05 | 15.48 | 19.65 | 23.36 | 27.08 | 30.80 | 57.02 | 57.02 | 57.02 |
| | New | - | 3.88 | 8.05 | 15.48 | 19.65 | 23.36 | 27.08 | 30.80 | 57.02 | 57.02 | 57.02 | - | - |

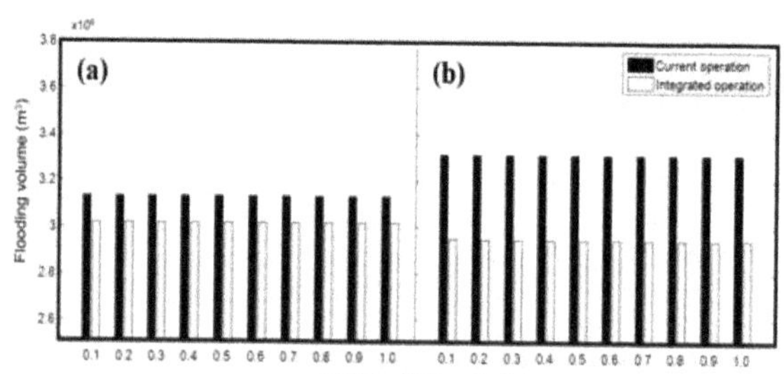


**Figure 11.** The results of current and integrated operations: (**a**) 2010; (**b**) 2011.

Overall, the integrated operation demonstrated good results, although they varied slightly according to the operating levels. When the level of the monitoring node was 0.2D, the maximum flooding volume was 2,743,103 $m^3$, and the minimum flooding volume was 2,741,478 $m^3$ with a level of 0.3D. Figure 11b shows the results of the current and integrated operations for 2011. The results in Figure 11b also show that the integrated operation produced a lower flooding volume than the current operation (3,312,733 $m^3$) for all levels of the monitoring node. The integrated operation once more showed good results, although they still slightly differed from one other according to the operating levels. The maximum flooding volume was 2,951,973 $m^3$ when the level of the monitoring node was 0.1D, and the minimum flooding volume was 2,944,196 $m^3$ when the level was 0.9D. The results of

Figure 11 show that the integrated operation was steadily better than the current operation at all levels of the monitoring nodes.

In Figure 12, the results of flooding volume over time using the integrated operation and current operation are compared. The results in 2010 and 2011 are shown in Figure 12a,b, respectively. The results in 2010 and 2011 show that the integrated operation had less flood volume per minute than the current operation.

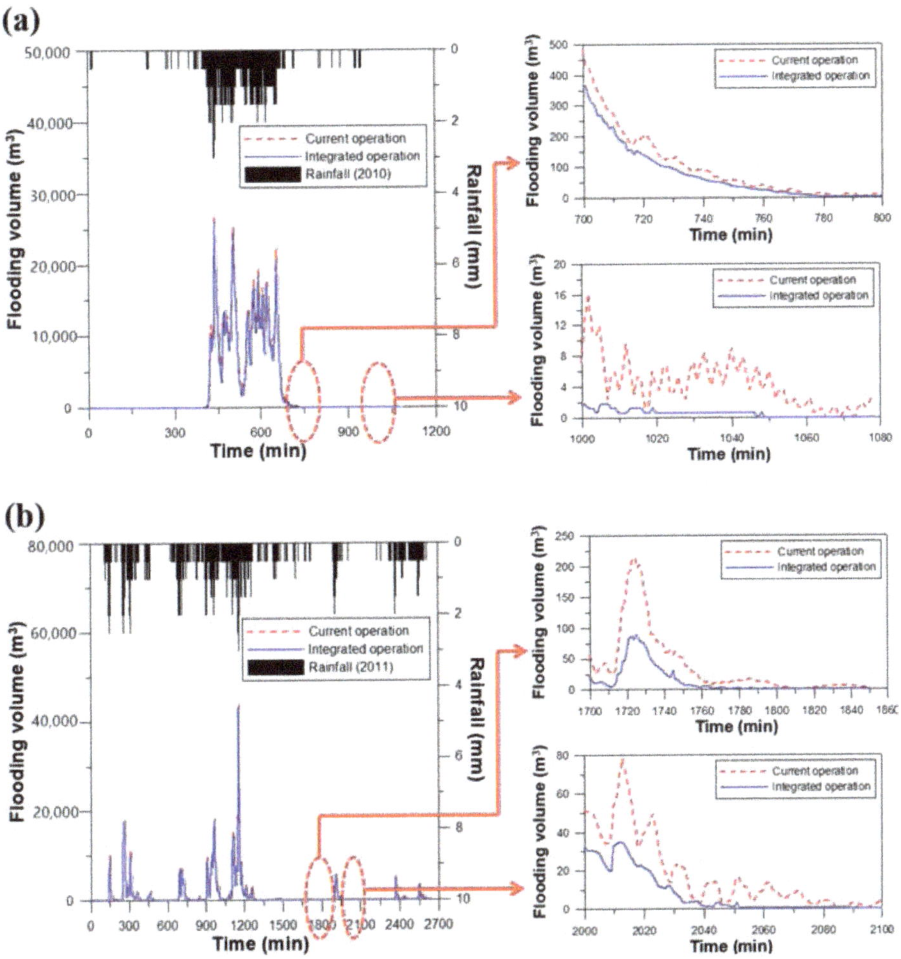


**Figure 12.** Rainfall and flooding volume over time: (**a**) 2010; (**b**) 2011.

Moreover, this study applied the system resilience to verify the ability to prepare for and recover from the malfunction (failure) of drainage facilities and inundation (system degradation) of drainage systems. The proposed resilience index was applied to the current and integrated operations for the 2010 and 2011 events. The results of the two operations were compared for both years, when the level of the monitoring node at the beginning of the integrated operation (including the CR operation) was 0.8D. The results of system resilience for the current and integrated operations are shown in Table 10.

**Table 10.** The results of system resilience for current and integrated operations.

| Event | System Resilience of the Current Operation | System Resilience of the Integrated Operation | Increment in System Resilience |
|---|---|---|---|
| 2010 | 0.199 | 0.238 | 0.039 |
| 2011 | 0.064 | 0.235 | 0.171 |

For the 2010 event, the system resilience of the current operation was 0.199, whereas that of the integrated operation was 0.238. This means that the integrated operation increases the system resilience of the urban drainage system in the target watershed by 0.039. The value of system resilience ranges from 0 to 1. A high value of system resilience in urban drainage systems means a more resilient drainage system. The integrated operation makes the urban drainage system in the target watershed resilient to failure (flooding).

For the 2011 event, the system resilience of the current operation was 0.064, whereas that of the integrated operation was 0.235. The system resilience increment between the current and integrated operations was thus 0.171. The system resilience of the current operation for the 2011 event was lower than that for 2010 because the total flooding volume of the 2011 event was larger than that in 2010 and system failure occurred frequently in 2011, as the flooding volume was widely distributed. The system resilience of the target watershed in Table 9 was calculated as low because the value of performance was calculated as zero if flooding occurred when there was no rainfall.

## 4. Conclusions

1. The first highlight in this study is the difference between the current operation and the integrated operation in urban streams with pump stations/decentralized reservoirs. The current operation is based on the vertical/one-way network and the integrated operation is based on the horizontal/interactive network. In the integrated operation, the operation of one drainage facility needs the information of the urban stream, and other drainage facilities also require the information of the urban stream. Furthermore, the urban stream is affected by the discharge of upstream drainage facilities. Therefore, information of all drainage facilities in a target area should be collected in the integrated operation with a horizontal/interactive network.
2. The second highlight is the effect of the integrated operation. The integrated operation was applied to two historical events (2010, 2011). The duration of the 2010 event (1000 min) is shorter than that of the 2011 event (2500 min). As the results of the flooding volume and system resilience indicate, the effect of integrated operation in 2011 is greater than that in 2011. It means that the integrated operation can be more effective in a rainfall event with a long duration. Consequently, the effect of the integrated operation can vary greatly depending on the operating time.
3. The third highlight is the system resilience by previous resilience indices and the revised resilience index. The system resilience in the revised resilience index in both operations is relatively low because all values are not over 0.3. The absolute difference of system resilience in both operations is small, but the relative difference is improved because the improvement of system resilience in the 2010 and 2011 events is 20% and 267%, respectively.
4. Finally, the results of this study will enable the evaluation of urban watershed management and the application of structural and nonstructural measures in urban areas. The integrated operation constitutes a nonstructural measure for reducing urban inundation and managing drainage facilities by considering the status of urban streams. The revised resilience index will be useful in establishing a policy regarding structural and nonstructural measures, such as the installation and operation of drainage facilities in urban areas. The integrated operation approach and revised resilience index will also help in the construction of a resilient city, which has recently attracted attention. Future research will be required regarding the operation of drainage facilities in metropolitan cities and the evaluation of flooding damage.

**Author Contributions:** E.H.L. carried out the survey of previous studies. E.H.L. and Y.H.C. wrote the original manuscript. E.H.L. conducted the simulations. Y.H.C. revised the original manuscript. E.H.L., Y.H.C., and J.H.K. conceived the original idea of the proposed method.

**Funding:** This research was funded by the National Research Foundation (NRF) of Korea in the Korean government (MISP) (No. 2016R1A2A1A05005306).

**Acknowledgments:** This work was supported by a grant from The National Research Foundation (NRF) of Korea in the Korean government (MSIP) (No. 2016R1A2A1A05005306).

**Conflicts of Interest:** The authors declare no conflict of interest.

## References

1. Galelli, S.; Goedbloed, A.; Schwanenberg, D.; van Overloop, P.J. Optimal real-time operation of multipurpose urban reservoirs: Case study in Singapore. *J. Water Res. Plan.-ASCE* **2012**, *140*, 511–523. [CrossRef]
2. Zacharof, A.I.; Butler, D.; Schütze, M.; Beck, M.B. Screening for real-time control potential of urban wastewater systems. *J. Hydrol.* **2004**, *299*, 349–362. [CrossRef]
3. Beeneken, T.; Erbe, V.; Messmer, A.; Reder, C.; Rohlfing, R.; Scheer, M.; Schuetze, M.; Schumacher, B.; Weilandt, M.; Weyand, M. Real time control (RTC) of urban drainage systems—A discussion of the additional efforts compared to conventionally operated systems. *Urban Water J.* **2013**, *10*, 293–299. [CrossRef]
4. Cembrano, G.; Quevedo, J.; Salamero, M.; Puig, V.; Figueras, J.; Martı, J. Optimal control of urban drainage systems. A case study. *Control Eng. Pract.* **2004**, *12*, 1–9. [CrossRef]
5. Hsu, N.S.; Huang, C.L.; Wei, C.C. Intelligent real-time operation of a pumping station for an urban drainage system. *J. Hydrol.* **2013**, *489*, 85–97. [CrossRef]
6. Schütze, M.; Campisano, A.; Colas, H.; Schilling, W.; Vanrolleghem, P.A. Real time control of urban wastewater systems—where do we stand today? *J. Hydrol.* **2004**, *299*, 335–348. [CrossRef]
7. Xu, W.D.; Fletcher, T.D.; Duncan, H.P.; Bergmann, D.J.; Breman, J.; Burns, M.J. Improving the multi-objective performance of rainwater harvesting systems using real-time control technology. *Water* **2018**, *10*, 147. [CrossRef]
8. Raimondi, A.; Becciu, G. On pre-filling probability of flood control detention facilities. *Urban Water J.* **2015**, *12*, 344–351. [CrossRef]
9. Sweetapple, C.; Astaraie-Imani, M.; Butler, D. Design and operation of urban wastewater systems considering reliability, risk and resilience. *Water Res.* **2018**, *147*, 1–12. [CrossRef] [PubMed]
10. Lee, E.H.; Lee, Y.S.; Joo, J.G.; Jung, D.; Kim, J.H. Investigating the impact of proactive pump operation and capacity expansion on urban drainage system resilience. *J. Water Resour. Plan. Manag.* **2017**, *143*, 04017024. [CrossRef]
11. Tamoto, N.; Endo, J.; Yoshimoto, K.; Yoshida, T.; Sakakibara, T. Forecast-based operation method in minimizing flood damage in urban area. In Proceedings of the 11th International Conference on Urban Drainage, Edinburgh, UK, 31 August–5 September 2008; Volume 31.
12. Lund, N.S.V.; Falk, A.K.V.; Borup, M.; Madsen, H.; Steen Mikkelsen, P. Model predictive control of urban drainage systems: A review and perspective towards smart real-time water management. *Crit. Rev. Environ. Sci. Technol.* **2018**, *48*, 1–61. [CrossRef]
13. Fiorelli, D.; Schutz, G.; Klepiszewski, K.; Regneri, M.; Seiffert, S. Optimised real time operation of a sewer network using a multi-goal objective function. *Urban Water J.* **2013**, *10*, 342–353. [CrossRef]
14. Hsu, M.H.; Chen, S.H.; Chang, T.J. Inundation simulation for urban drainage basin with storm sewer system. *J. Hydrol.* **2000**, *234*, 21–37. [CrossRef]
15. Kroll, S.; Fenu, A.; Wambecq, T.; Weemaes, M.; Van Impe, J.; Willems, P. Energy optimization of the urban drainage system by integrated real-time control during wet and dry weather conditions. *Urban Water J.* **2018**, *15*, 1–9. [CrossRef]
16. Lee, E.H.; Lee, Y.S.; Joo, J.G.; Jung, D.; Kim, J.H. Flood reduction in urban drainage systems: Cooperative operation of centralized and decentralized reservoirs. *Water* **2016**, *8*, 469. [CrossRef]
17. Pleau, M.; Colas, H.; Lavallée, P.; Pelletier, G.; Bonin, R. Global optimal real-time control of the Quebec urban drainage system. *Environ. Modell. Softw.* **2005**, *20*, 401–413. [CrossRef]
18. Vanrolleghem, P.A.; Benedetti, L.; Meirlaen, J. Modelling and real-time control of the integrated urban wastewater system. *Environ. Modell. Softw.* **2005**, *20*, 427–442. [CrossRef]

19. Haughton, G. Developing sustainable urban development models. *Cities* **2007**, *14*, 189–195. [CrossRef]
20. Chiesura, A. The role of urban parks for the sustainable city. *Landsc. Urban Plan.* **2004**, *68*, 129–138. [CrossRef]
21. Pickett, S.T.; Cadenasso, M.L.; Grove, J.M. Resilient cities: Meaning, models, and metaphor for integrating the ecological, socio-economic, and planning realms. *Landsc. Urban Plan.* **2004**, *69*, 369–384. [CrossRef]
22. Egger, S. Determining a sustainable city model. *Environ. Modell. Softw.* **2006**, *21*, 1235–1246. [CrossRef]
23. Blackmore, J.M.; Plant, R.A. Risk and resilience to enhance sustainability with application to urban water systems. *J. Water Resour. Plan. Manag.* **2008**, *134*, 224–233. [CrossRef]
24. Butler, D.; Farmani, R.; Fu, G.; Ward, S.; Diao, K.; Astaraie-Imani, M. A new approach to urban water management: Safe and sure. *Procedia Eng.* **2014**, *89*, 347–354. [CrossRef]
25. Mugume, S.N.; Gomez, D.E.; Fu, G.; Farmani, R.; Butler, D. A global analysis approach for investigating structural resilience in urban drainage systems. *Water Res.* **2015**, *81*, 15–26. [CrossRef] [PubMed]
26. Siekmann, T.; Siekmann, M. Resilient urban drainage—Options of an optimized area-management. *Urban Water J.* **2015**, *12*, 44–51. [CrossRef]
27. Mugume, S.N.; Butler, D. Evaluation of functional resilience in urban drainage and flood management systems using a global analysis approach. *Urban Water J.* **2016**, *14*, 1–10. [CrossRef]
28. Todini, E. Looped water distribution networks design using a resilience index based heuristic approach. *Urban Water J.* **2000**, *2*, 115–122.
29. Bruneau, M.; Chang, S.E.; Eguchi, R.T.; Lee, G.C.; O'Rourke, T.D.; Reinhorn, A.M.; Shinozuka, M.; Tierney, K.; Wallace, W.A.; Winterfeldt, D. A framework to quantitatively assess and enhance the seismic resilience of communities. *Earthq. Spectra.* **2003**, *19*, 733–752. [CrossRef]
30. Godshalk, D. Urban hazard mitigation: Creating resilient cities. *Nat. Hazards Rev.* **2003**, *4*, 136–143. [CrossRef]
31. Ministry of Land, Infrastructure, and Transport of the Korean government. *River Management Plan for Dorim Stream*; Ministry of Land, Infrastructure, and Transport of the Korean government: Seoul, Korea, 2002.
32. United States Environmental Protection Agency. *Storm Water Management Model user's Manual Version 5.0. EPA*; United States Environmental Protection Agency: Washington, DC, USA, 2010.
33. Seoul Metropolitan Government. *Report on Design and Expansion of Guro1 Pump Station*; Seoul Metropolitan Government: Seoul, Korea, 2010.
34. Seoul Metropolitan Government. *Report on Design and Expansion of Guro2 Pump Station*; Seoul Metropolitan Government: Seoul, Korea, 2010.
35. Seoul Metropolitan Government. *Report on Design and Expansion of Guro3 Pump Station*; Seoul Metropolitan Government: Seoul, Korea, 2010.
36. Seoul Metropolitan Government. *Report on Design and Expansion of Guro4 Pump Station*; Seoul Metropolitan Government: Seoul, Korea, 2010.
37. Seoul Metropolitan Government. *Report on Design and Expansion of Mullae Pump Station*; Seoul Metropolitan Government: Seoul, Korea, 2010.
38. Seoul Metropolitan Government. *Report on Design and Expansion of Dorim2 Pump Station*; Seoul Metropolitan Government: Seoul, Korea, 2010.
39. Seoul Metropolitan Government. *Report on Design of Daerim2 Pump Station*; Seoul Metropolitan Government: Seoul, Korea, 2010.
40. Seoul Metropolitan Government. *Report on Design and Expansion of Daerim3 Pump Station*; Seoul Metropolitan Government: Seoul, Korea, 2010.
41. Seoul Metropolitan Government. *Report on Design and Expansion of Sinlim1 Pump Station*; Seoul Metropolitan Government: Seoul, Korea, 2012.
42. Seoul Metropolitan Government. *Report on Design of Sinlim2 Pump Station*; Seoul Metropolitan Government: Seoul, Korea, 2012.
43. Seoul Metropolitan Government. *Report on Design of Sinlim5 Pump Station*; Seoul Metropolitan Government: Seoul, Korea, 2012.
44. Seoul Metropolitan Government. *Report on Design of Gwanak Detention Reservoir*; Seoul Metropolitan Government: Seoul, Korea, 2013.
45. Yeongdeungpo-Gu. *Report on Design of Daerim Detention Reservoir*; Yeongdeungpo-Gu: Seoul, Korea, 2007.
46. Ministry of Public Safety and Security. *The Disaster Year Book*; Ministry of Public Safety and Security: Seoul, Korea, 2010.

47. Ministry of Public Safety and Security. *The Disaster Year Book*; Ministry of Public Safety and Security: Seoul, Korea, 2011.
48. Huff, F.A. Time distribution of rainfall in heavy storms. *Water Resour. Res.* **1967**, *3*, 1007–1019. [CrossRef]
49. Yoon, Y.N.; Jung, J.H.; Ryu, J.H. Introduction of design flood estimation. *J. Korea Water Resour. Assoc.* **2013**, *46*, 55–68.
50. Korea Precipitation Frequency Data Server. Available online: www.k-idf.re.kr (accessed on 27 May 2018).
51. Lee, E.H.; Kim, J.H. Development of a flood-damage-based flood forecasting technique. *J. Hydrol.* **2018**, *563*, 181–194. [CrossRef]

*water*

MDPI

*Article*

# Residential Flood Loss Assessment and Risk Mapping from High-Resolution Simulation

**Zulfahmi Afifi [1], Hone-Jay Chu [2,*], Yen-Lien Kuo [3], Yung-Chia Hsu [4], Hock-Kiet Wong [5] and Muhammad Zeeshan Ali [2]**

1 International Master Program on Natural Hazards Mitigation and Management, National Cheng Kung University, Tainan City 701, Taiwan; nc6067016@mail.ncku.edu.tw

2 Department of Geomatics, National Cheng Kung University, Tainan City 701, Taiwan; Zeeshanktk1992@yahoo.com

3 Department of Economics, National Cheng Kung University, Tainan City 701, Taiwan; yenlien@mail.ncku.edu.tw

4 Hydrotech Research Institute, National Taiwan University, Taipei 10617, Taiwan; ychsu1978@gmail.com

5 Department of Hydraulics and Ocean Engineering, National Cheng Kung University, Tainan City 701, Taiwan; ikhowong93@hotmail.my

* Correspondence: honejaychu@gmail.com

Received: 12 March 2019; Accepted: 8 April 2019; Published: 10 April 2019

**Abstract:** Since the patterns of residential buildings in the urban area are small-sized and dispersed, this study proposes a high-resolution flood loss and risk assessment model to analyze the direct loss and risk impacts caused by floods. The flood inundation simulation with a fine digital elevation model (DEM) provides detailed estimations of flood-inundated areas and their corresponding inundation depths during the 2016 Typhoon Megi and 2017 Typhoon Haitang. The flood loss assessment identifies the impacts of both events on residential areas. The depth-damage table from surveys in the impacted area was applied. Results indicated that the flood simulation with the depth-damage table is an effective way to assess the direct loss of a flood disaster. The study also showed the effects of spatial resolution on the residential loss. The results indicated that the low-resolution model easily caused the estimated error of loss in dispersed residential areas when compared with the high-resolution model. The analytic hierarchy process (AHP), as a multi-criteria decision-making method, was used to identify the weight factor for each vulnerability factor. The flood-vulnerable area was mapped using natural and social vulnerability factors, such as high-resolution DEM, distance to river, distance to fire station, and population density. Eventually, the flood risk map was derived from the vulnerability and flood hazard maps to present the risk level of the flood disaster in the residential areas.

**Keywords:** flood; 3Di; loss assessment; analytic hierarchy process (AHP); risk map

## 1. Introduction

Floods are some of the most catastrophic natural disasters that include severe economic impacts, especially if floods happen in big cities around the world. In Taiwan, floods in the last 25 years have resulted in losses of $518 million USD. The flood losses are approximately 4.6 times the losses caused by fire damage [1]. Floods result in the exposure of many properties in residential areas to standing water. Inundation may cause structural damages, such as on wall linings, and property damage, including electronic devices [2]. Flood damages and losses can be broadly divided into two types: direct losses and indirect losses. Direct flood losses are those caused by the physical or structural impact of a flood event while indirect losses are not the direct economic losses of a flood. In the majority of damage estimation studies, losses are restricted to direct losses for each element [3]. The evaluation of flood loss is related to the use of damage functions from potential asset damaging [4]. However, flood loss is highly correlated to flood water depth. The depth-damage curve or table was created via a field survey of direct loss in a flood event in the flood loss model [5]. The loss assessment includes asset exposure, susceptibility to suffer damage, and damaging potential [6]. Moreover, flood risk was usually defined as the product of hazard e.g., probability of occurrence and vulnerability or estimated cost of foreseeable damage [7]. Risk assessment was generally measured as the expected loss degree of risk as a consequence of a hazardous event [8].

Flood loss and risk assessment have been determined through flood hazard data, i.e., spatial distribution of flood area, exceedance probability, and asset vulnerability data [6]. However, most flood information is only based on point discharge measurements at discrete locations without spatiotemporal information of flood inundation [9]. For the lack of sufficient depth measurements, we offer a flood loss and risk assessment that integrates 2D hydraulic model results and high-resolution digital elevation model (DEM) data, especially during the typhoon events. Considering the pattern of residential area in the urban area, the water depth of each residential area is more appropriate based on the high-resolution inundation simulation. A simulation-based flood loss and risk assessment can provide an effective way for the spatial estimation and visualization of flood loss and risk in potential flooding areas. Eventually, the impact of estimated loss or risk using a geographic information system (GIS) is practical for flood loss assessment and disaster management [4,5,10].

This study proposes a GIS-based high-resolution flood loss and risk assessment model to analyze the direct loss and risk impacts caused by a flood in a residential urban area. The model with fine grids provides a high-resolution estimation of a flood-inundated area and its corresponding loss and risk. The flood simulated water depths based on typhoon events during the 2016 Typhoon Megi and 2017 Typhoon Haitang were used in this research. The flood loss assessment identifies the impact of a simulated flood water depth in the residential area. The highest water depths in the simulation of the typhoon event were used to estimate the maximum loss in residential buildings. The depth-damage table of the impacted area was applied in our survey for flood loss assessment. In addition, risk is defined as the product of hazard e.g., high-resolution water depth and flood vulnerability. The flood-vulnerable area was mapped using natural and social vulnerability factors, such as distance from the river, distance to the fire station, elevation, and people density. The analytic hierarchy process (AHP) was used to acquire the weight factor of each vulnerability factor. A flood risk map was derived from the vulnerability and flood hazard maps to show the risk of the flood disaster.

## 2. Datasets and Methods

Figure 1 describes the process of each element in different colors. Each component represents the proposed material and method used to derive the results. The black and blue colors indicate the materials used in the beginning of the study. The current land use map was used to analyze the impact of the flood disasters on the residential area. The high-resolution (1 m resolution) DEM was used in this study. Water depth data were generated from the 3Di model [11]. The DEM, distance to river, distance to fire station, and population density were used as natural and social vulnerability factors. The green color indicates the process of deriving the flood loss and hazard map. The orange color indicates the process of generating the vulnerability map. The blue color indicates the process of deriving the risk map as the final step. A GIS model was also used to make the flood loss and risk process automatic. This model combined loss and risk functions and processes developed using Python. The proposed model enhances the applications with the purpose of flood loss and risk assessment by combining functions and data related to flood loss, hazard, and vulnerability (Figure 1).

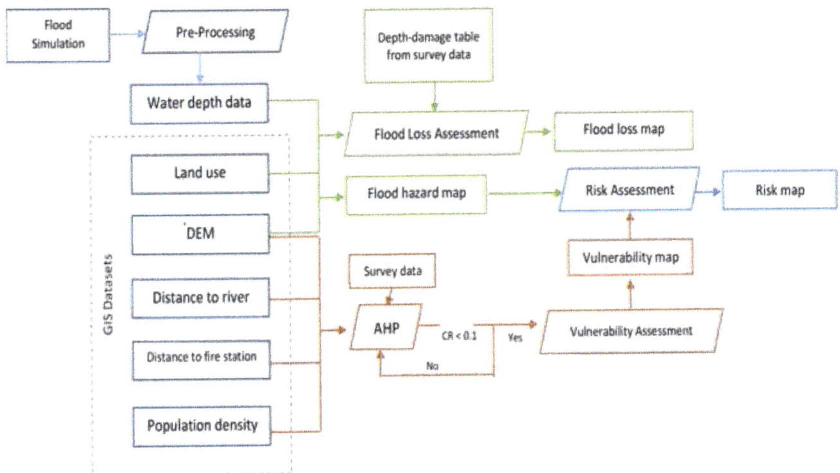


**Figure 1.** Flowchart of the GIS-based flood loss and risk assessment.

### *2.1. Flood Model*

The 3Di model was developed by Professor Stelling from TU Delft, Netherlands in 2010 [11]. The 3Di model combines four numerical methods including the subgrid method, bottom friction based on the concepts of roughness depth, the staggered-grid finite-volume method for shallow water equations with rapidly varying flows, and the quad-tree technique [12]. The 3Di model with sub-grid and quad-tree methods can handle a large number of computational grids with high-resolution topographic data [13]. The 3Di numerical method utilizes a finite volume. The method works by solving the 2D shallow water equations and combining the data through the quad-tree method. Such process increases the density of computational cells in important areas, such as urban areas, levees, and roads. It also ensures that the flow field will perform smoothly and reduces errors affected by the terrains. 3Di can efficiently perform under high-resolution terrain data. The integrated 3Di model consists of two main components including 2D and 1D-flow components. In the model simulation, 3Di uses 1D and 2D equations in terms of terrain conditions.

Compared with other inundation models, the most special part is the technique of the simulation cell in the 3Di model, such as the sub-grid and quad-tree methods. The sub-grid method works before starting a simulation, and it provides a series of temporary tables that contain the water volume–level relationship of each simulation cell and link during the simulation when a water level is being solved. Accordingly, we determine the volume of water that has changed between two adjacent cells. The quad-tree method is suitable for simulating high-resolution terrain data or a large simulation area. It has been widely used in inundation simulation or hydraulic calculation. This method can increase the flexibility of simulation cells and is efficient when used for simulation.

In this study, the high-resolution simulation model was based on the high-resolution DEM. The distributed rainfall data used in the 3Di flood simulation data were the Quantitative Precipitation Estimation and Segregation Using Multiple Sensors (QPESUMS). QPESUMS is the rainfall time series developed by the Central Weather Bureau, the Soil and Water Conservation Bureau, and the Water Resources Agency in 2002. This time series integrates radar, satellite, rain gauge station, and lightning observation data and is combined with the QPESUMS development. The size of QPESUMS rainfall data is 1.3 km × 1.3 km, and it collects rainfall data every 10 minutes. Observed water level data from two stations were used to validate the flood model data result. The gauging stations of the Wangliao Bridge and the Yanzi Bridge are located in the Yongkang District. The water level data will be updated every 10 min. Moreover, the simulated water levels and depths from the 3Di are the vector data in the form of quad-tree data of the water level in the study area. These data were then converted into raster data for the subsequent loss and risk assessment. The flood raster data were converted to different resolutions to determine the effects of differences in the resolution. The flood losses in using different spatial resolutions (1, 5, and 10 m) were compared in this study.

### 2.2. Loss Assessment

The flood loss assessment was calculated based on a direct assessment method of the impact of flooding on residential buildings. This assessment used the maximum water depth data of the flood simulation model from the 3Di model during the flood events and the depth-damage table from the field survey.

The depth-damage table was generated via the field survey in the flood area in the Rende District using the synthetic method [14]. Fourteen local residents joined the loss survey. The survey was conducted using a questionnaire given to residents in the flood disaster-prone areas. The table was generated from the survey results of the flooding areas, depths, and the losses in the residential area including furniture, clothes, electronic devices, and domestic appliances [14]. The survey data implied that no costs were incurred by the community in the disaster-prone area for the flood recovery work. The building recovery work and cleaning process were performed by the homeowners themselves and assisted by the government, such as firefighters and military personnel.

### 2.3. Risk Assessment

A flood risk map is useful for increasing the awareness of local communities, local authorities, and other organizations on flood risks. The determination of flood-risk areas is normally accomplished through a subjective method using vulnerability data and the water depth of flood, with the concept of risk itself interpreted in various ways. With a combination of the data hazard and vulnerability, the flood risk map was generated using automatic procedures. The flood risk was generated by multiplying the flood hazard map with the vulnerability map [15,16].

$$\mathrm{R} = \mathrm{H} \times \mathrm{V}, \tag{1}$$

where R is the risk map, H is the flood hazard map, and V is the flood vulnerability map. The maximum water depth of each typhoon event was determined from the flood simulation. Furthermore, the average water depth from two events were applied for the flood hazard map. In this study, the five-level risk map was generated for flood risk assessment.

### 2.4. Vulnerability Assessment

Flood vulnerability mapping is the process of determining the degree of susceptibility of a given place to flooding. Vulnerability is the most crucial component of risk since it determines whether exposure to a hazard constitutes a risk that may actually result in a disaster. Flood vulnerability can be obtained from the weighted product model. This flood vulnerability map will be combined with available vulnerability data in the study area.

$$\mathrm{V} = \Pi_{\mathrm{i}=1}^{\mathrm{n}} {\mathrm{V}_\mathrm{i}}^{\mathrm{w}_\mathrm{i}}, \tag{2}$$

where $\mathrm{w_i}$ is the weight of the factor i (i.e., DEM, distance to river, distance to fire station, and population density) from the AHP. $\mathrm{V_i}$ is the vulnerability map for factor i and n is the number of factors. The final result is a flood vulnerability map that has information on vulnerability levels ranging from very low to extreme levels.

The vulnerability factor data in this study include natural and social factors, such as the DEM, distance to river, distance to fire station, and population density. In Tainan, low-lying areas have a high vulnerability to the dangers of floods. The DEM data were used as one of the vulnerability factors in this study. Flood happens when rainwater exceeds the capacity of waterways or rivers. Distance to the river is also an important factor in the analysis of vulnerability to flooding. River line area data were also included to calculate vulnerability in terms of the distance of an area to river flow. Areas that are close to the river will tend to have a higher value of vulnerability than other areas located far away from the river. Distance to the river is determined through GIS. Moreover, fire stations in Tainan are considered the first agency that can handle emergencies during a disaster, including floods. A region that has a distance far from a fire station will have a higher vulnerability when a flood occurs than an area located close to the fire station. This vulnerability is interpreted as the first response to an emergency situation to prevent casualties and dangers that can cause losses due to flooding. People can acquire help from firefighters and military services. Thus, fire stations are critical facilities when considering the social vulnerability index [17]. In addition, population density is an important factor for vulnerability that must be considered. The Ministry of Interior Affairs provides the village-based population data. The polygon data are converted to a raster dataset through GIS.

### 2.5. AHP for Vulnerability Weight

The flood disaster vulnerability map was created by processing some of the available vulnerability factor data. A multi-criterion decision-making (MCDM) method was used to determine the importance of vulnerability factors from the vulnerability factor data. One of the MCDM methods used was the AHP. AHP is a semi-quantitative decision-making value judgment approach that fulfills the objectives of decision makers [18]. In this study, the AHP model [19] was used. The AHP continued until the obtained consistency ratio was less than 0.1.

Twenty participants were selected for the AHP survey. Among them, 40% were local students from the International Master Program on Natural Hazard Mitigation and Management Program, 50% were hydraulic engineering students, and 10% were social science students. The backgrounds of the participants were acquired to examine their various perspectives.

### 2.6. Study Area

The study area is Tainan City in Taiwan, which covers an area of 137 km$^2$. The locations of the study area (blue polygon) and residential area (orange polygon) are shown in Figure 2. The river (Erren River) moves through the middle of the city. Typhoons, e.g., the 2016 Typhoon Megi and 2017 Typhoon Haitang, were accompanied by abundant rainfall (352 and 430 mm with 24-hour accumulated rainfall) that caused serious damages in Tainan.

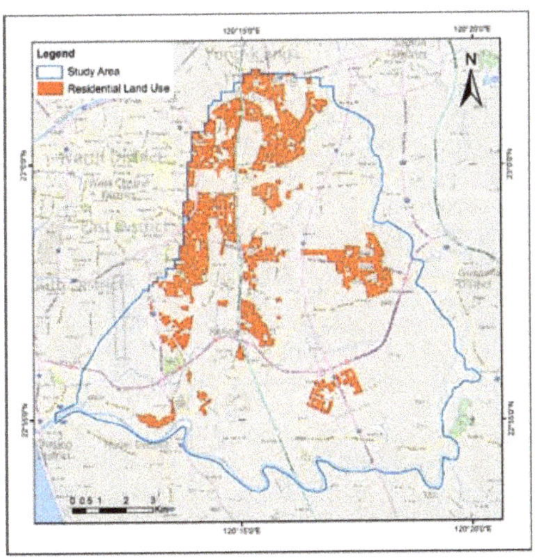


**Figure 2.** Study area and residential area.

## 3. Result and Discussion

### 3.1. Flood Hazard Map, Depth-Damage Table, and Vulnerability Weight by AHP

A flood hazard map was created on the basis of the data of the flood simulation model that was processed with an automatic GIS model. The flood simulation data were obtained from the research generated from the 3Di model. Water depth data from two flood events were used in this study (the 2016 Typhoon Megi and the 2017 Typhoon Haitang). The average water depths from both typhoon events were used to generate the flood hazard maps. Water level data from two stations known as the Wangliao Bridge and the Yanzi Bridge were used to validate the flood model data result. Figure 3 shows the comparison between flood simulations and observation of the water level in the two typhoon events. The red color represents the observation of the water level and the blue one represents the simulation of the water level. The results showed a good fitting between the simulated and observed water levels. Figure 4 shows the maximum flood water depth of each typhoon for the flood hazard map. The intensity level of the blue color indicates water depth. The dark blue indicates the highest water level in the study area. The automatic model was developed based on the flood data simulation caused by the typhoon disaster. In the 2016 Typhoon Megi and 2017 Typhoon Haitang, the results showed that the maximum flood water level ranged from 0.25 m to 2 m in the entire study area.

Table 1 shows the depth-damage table generated from the field survey in the flood area in the Rende District. No damage losses occurred if the water depth was less than 0.15 m. The flood loss increased with water depth in the depth-damage table. The damage per area is at maximum if the water depth is higher than 2 m. Figure 5 shows the vulnerability factor data such as (a) DEM, (b) distance to river, (c) distance to fire station, and (d) population density. In Figure 5a, the elevation of the study area is between −0.5 m and 40 m. Figure 5b shows that the residential area is close to rivers. In Figure 5c, the fire station is uniformly distributed except in the southeast area. Figure 5d shows that the population density is less than three persons per 25 $m^2$ in most areas but is crowded with 10 persons to 59 persons per 25 $m^2$ in the area. The urban population is concentrated in the core area of the northern part. The vulnerability factor is an important element for minimizing the impact of floods in the future.

Table 2 shows the vulnerability weight value from the AHP. The results of the AHP method is the weight value used to determine the level of importance of the vulnerability factor. The two most important factors are the population density (weight = 0.310) and distance to river (weight = 0.271). Vulnerability weights contribute in the evaluation of flood risk and are sensitive due to their specific local factors [20]. AHP modeling can identify vulnerability on a local scale but is difficult to impeccably quantify because it depends on the quality of the indicators and survey data.

**Table 1.** Depth-damage table for the residential area from the field survey.

| Depth-Damage | Depth (m) | | | | |
|---|---|---|---|---|---|
| | <0.15 | 0.15–0.35 | 0.35–0.7 | 0.7–0.9 | ≥2 |
| Damage (NTD/$m^2$) | 0 | 242 | 1136 | 1623 | 3607 |

**Table 2.** Vulnerability weight values from the AHP.

| AHP Factor | Elevation | Distance to River | Distance to Fire Station | Population Density |
|---|---|---|---|---|
| Weight | 0.202 | 0.271 | 0.217 | 0.310 |

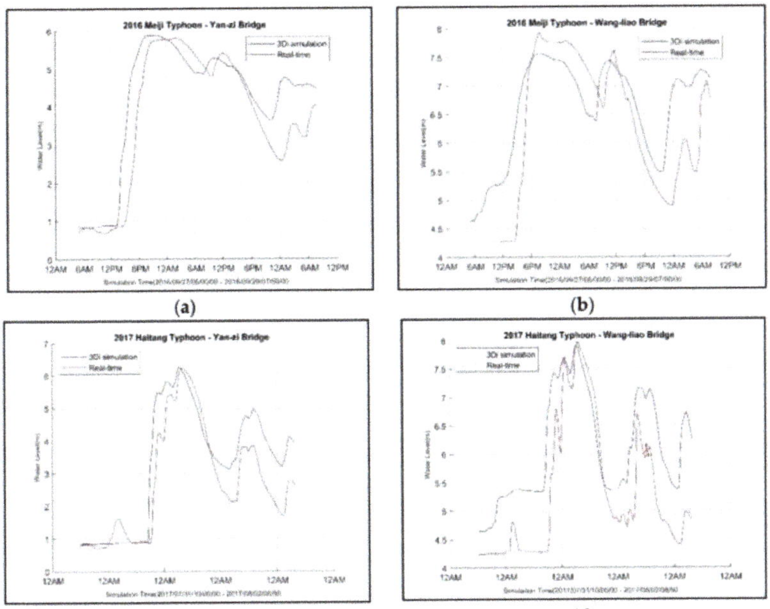


**Figure 3.** Comparison between flood simulation (blue color) and observation (red color) from the water level station at the (**a**) Yanzi Bridge during Typhoon Megi, (**b**) the Wangliao Bridge during Typhoon Megi, (**c**) the Yanzi Bridge during Typhoon Haitan, and (**d**) the Wangliao Bridge during Typhoon Haitang (unit: m).

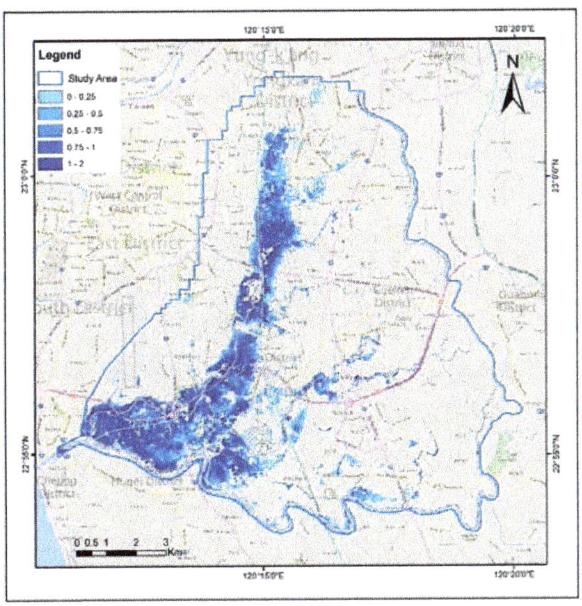


(**a**)

**Figure 4.** *Cont.*

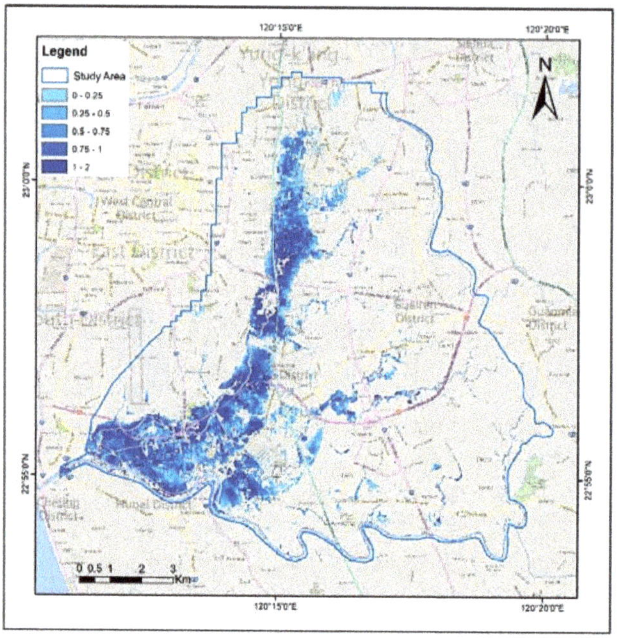


(**b**)

**Figure 4.** Flood hazard map: maximum flooding water depth during (**a**) 2016 Typhoon Megi and (**b**) 2017 Typhoon Haitang (unit: m).

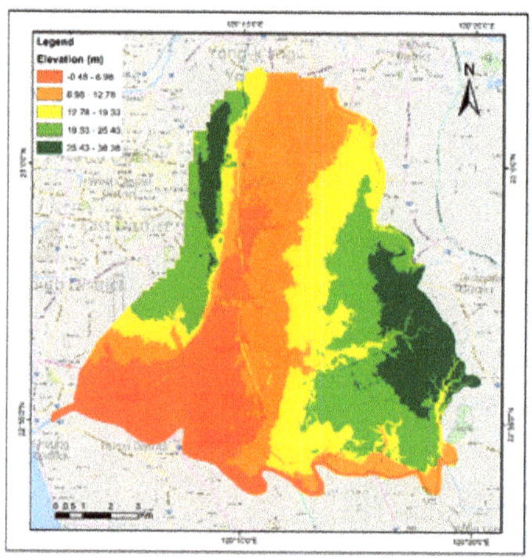


(**a**)

**Figure 5.** *Cont.*

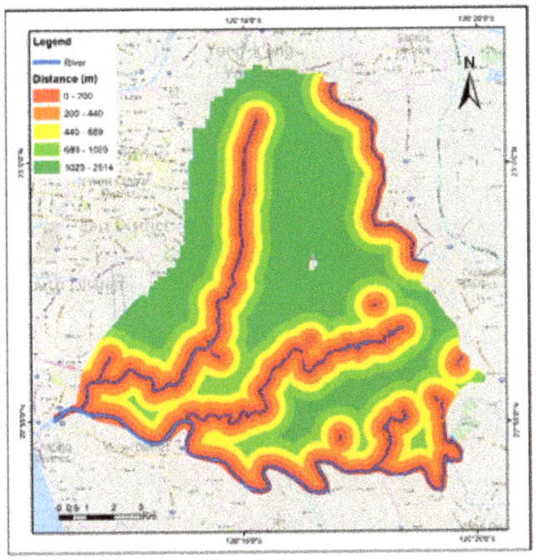


**(b)**

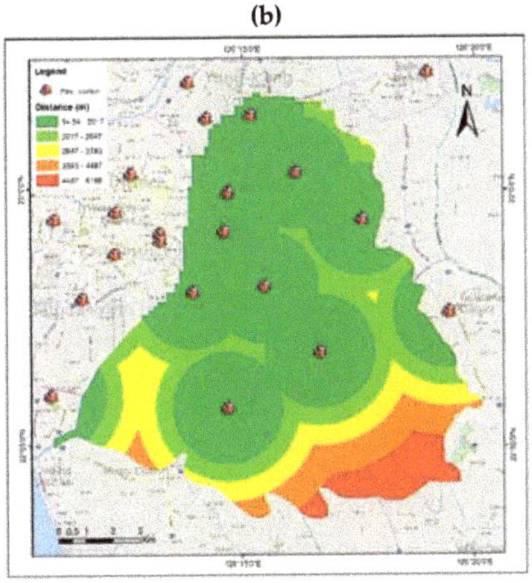


**(c)**

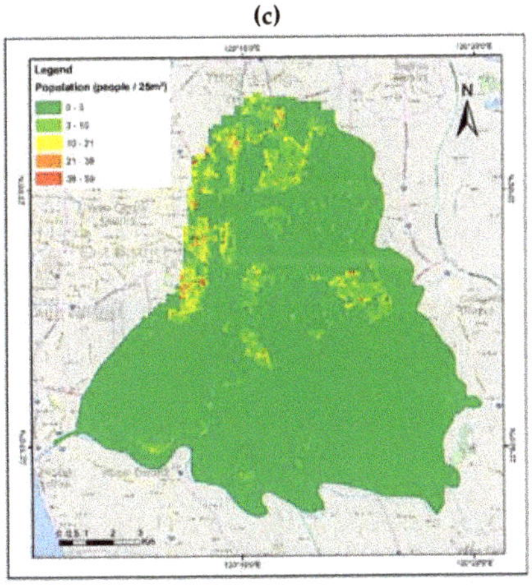


**(d)**

**Figure 5.** Vulnerability factors for (**a**) DEM (unit: m), (**b**) distance to river (unit: m), (**c**) distance to fire station (unit: m), and (**d**) population density (unit: persons/25 $m^2$).

### 3.2. Flood Loss Comparison Using Different Resolutions

In this subsection, the effects of different spatial model resolutions are discussed. Based on the loss comparisons with different spatial resolutions in Table 3, the estimated flood losses and flooding area at the fine resolutions of 1 m and 5 m resolutions were similar but were different from the 10 m resolution. Table 4 shows the comparison in flood loss at resolutions of 1 m to 5 m and 10 m. The loss result from the 10-m resolution contains a large difference among the cases. We explored the sensitivity of the flood loss estimates with the spatial representation in the flood loss model. The estimated loss at the 1 m or 5 m resolution is more appropriate due to the dispersed and small-sized residential buildings in Tainan City. Since the pattern of the residential area is dispersed, the high-resolution flood model offers relatively detailed loss assessment in the flood areas. This loss value will increase or decrease depending on the used scale of the study area. The case evaluated the sensitivity of the spatial resolution. In summary, the loss differences reached 7.8% and 11.2% during two events when compared with the 1-m and 10-m resolutions. Spatial resolutions used in the assessment had a direct impact on the potential flood heights [21]. However, the flood loss evaluation was often restricted due to the lack of dense observation data [22]. The adoption of a high-resolution flood simulation approach represents small-scale structural elements and small topographic variations [23]. The detailed simulation information is relevant for assessing the flood loss and risk in an urban area.

For flood management, the loss and risk are identified effectively based on simulation approaches [5,8]. The simulation model is applied to determine the maximum water depths during typhoon events without sufficient flood depth measurements. Since the patterns of residential areas in the study area are distributed, inundation depths in a residential area can be better identified using the high -resolution simulation model. However, the benefits of using the complex building representation and high-resolution flood inundation model are uncertain because of the lack of sufficient data for model calibration and validation [23].

**Table 3.** Flood losses from different spatial resolutions.

| Flood Event | Different Resolutions | | | | | |
|---|---|---|---|---|---|---|
| | 1 m | | 5 m | | 10 m | |
| | Area ($m^2$) | Residential Loss (NTD) | Area ($m^2$) | Residential Loss (NTD) | Area ($m^2$) | Residential Loss (NTD) |
| **Megi 2016** | 515,477 | 257,760,212 | 539,148 | 258,043,549 | 611,390 | 277,865,133 |
| **Haitang 2017** | 330,025 | 137,362,931 | 335,488 | 134,118,030 | 414,167 | 152,810,339 |

**Table 4.** Flood loss comparison of 1 m resolution.

| Flood Event | Residential Loss in Different Resolution (NTD) | | |
|---|---|---|---|
| | 1 m | 5 m | 10 m |
| **Megi 2016** | 257,760,212 | 258,043,549 | 277,865,133 |
| Difference from 1 m resolution | | 0.1% | 7.8% |
| **Haitang 2017** | 137,362,931 | 134,118,030 | 152,810,339 |
| Difference from 1 m resolution | | −2.4% | 11.2% |

### 3.3. Vulnerability and Risk Maps

A vulnerability map was derived from four vulnerability factors. Figure 6 shows the vulnerability results classified into five categories. Every vulnerability range is represented by a unique color. Level 1 (dark green color) indicates the areas with a very low vulnerability. Level 2 (light green) indicates the areas with low vulnerability, and level 3 (yellow color) indicates the areas with a moderate level of loss. Levels 4 and 5 (orange and red colors) indicate areas with high and very high vulnerabilities, respectively. The vulnerability map shows that the areas with the highest vulnerability are located in the south and center line because they are close to the river and are composed of lowland with high population density.

Figure 7 shows the risk map of the study area. The risk level is highly dependent on flood hazard and vulnerability. The flood risk is classified into the five levels of colors. Level 1 (dark green color) indicates the areas with very low risk. Level 2 (light green) indicates the areas with low risk, and level 3 (yellow color) indicates the areas with a moderate level of loss. Levels 4 and 5 (orange and red colors) indicate the areas with high and very high risks, respectively. In the risk map, most areas have no risk of flooding even though some have high vulnerability. The result also showed that the most areas at risk are located in the South District near the river area. The areas at risk of flooding are located around the river and have low elevation. The Tainan City Government can use the risk map as a reference to reduce the level of vulnerability in the areas around the river. Such a goal can be achieved by adding fire station units and reducing population density around the river. Reducing the value of this vulnerability factor will also have an impact for decreasing the risk of flood loss in the future.

The risk map analysis indicated that most areas (883,526 $m^2$) have a very low risk level, as denoted by the dark green color in Figure 7. A total of 269,192 $m^2$ areas are located in the low-risk area, as indicated by the light green color. Only 8099 $m^2$ areas are located in the moderate level of risk, as represented by the yellow color. In the flood risk maps, the southwestern parts of the district near the river were most vulnerable to floods. The risk value was calculated based on the water depth from the hazard map and the value of the vulnerability map. Most areas are not at risk for floods but some have high vulnerability. The government can reduce the risk value and reduce the social vulnerability value through population reduction and adding fire station in areas with high water depth value. Figure 8 identifies all residential areas that have risk values and very low risk level. Risk value is affected by the vulnerability factors and the hazard map in the risk-generating process. In Figure 8, no residential polygon that has a high level risk or a very high level risk was determined. Therefore, the residential polygon in the study area faces a low potential risk against flood disaster loss in the future. The government needs to pay attention to the areas, even those with less than level 3 of risk.

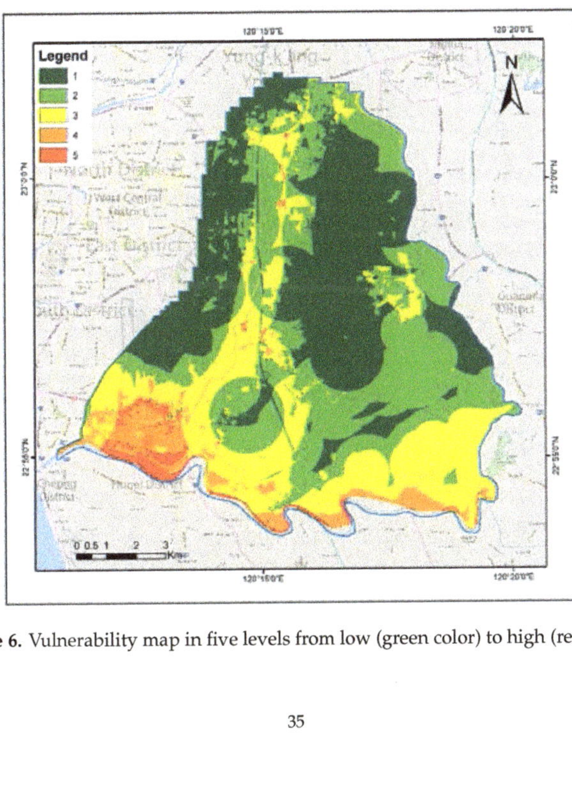


**Figure 6.** Vulnerability map in five levels from low (green color) to high (red color).

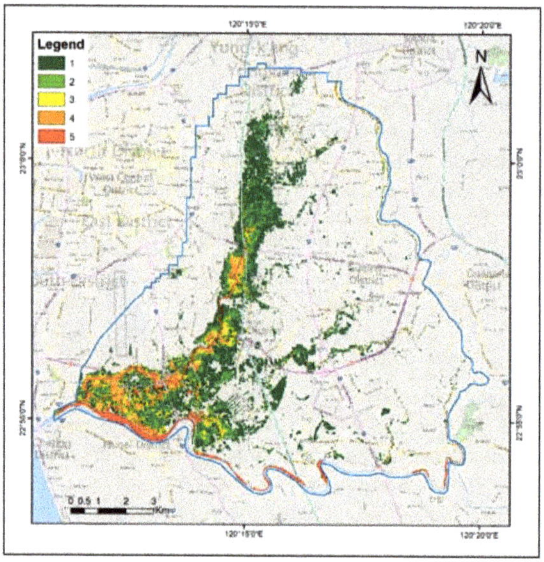


**Figure 7.** Risk map in five levels from low (green color) to high (red color).

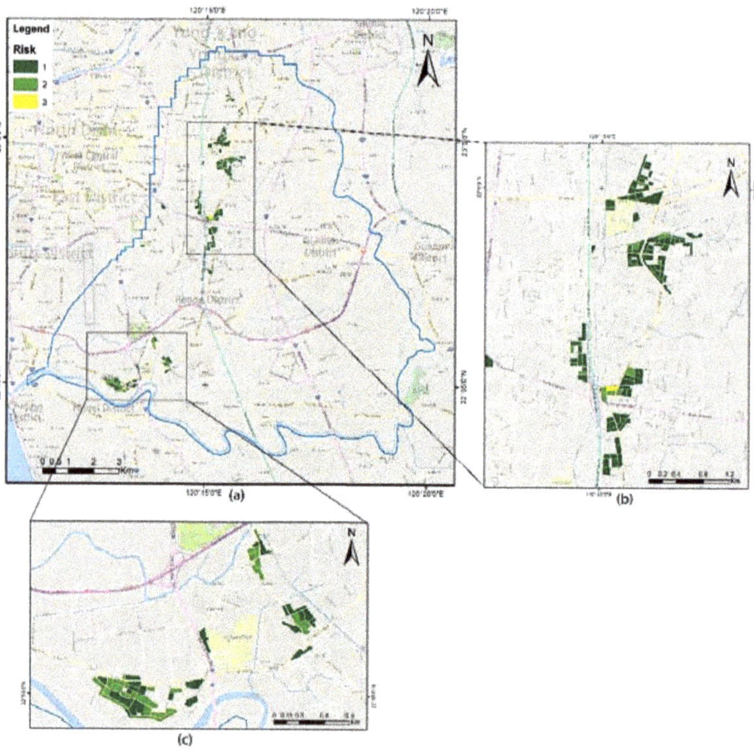


**Figure 8.** Risk map in a residential area: (**a**) risk in the entire study area, (**b**) details in the Rende District, and (**c**) details in the South District.

Flood risk is a product of the probability of the flood event occurrence and its consequence to society and the environment [24–26]. The 24-h accumulated rainfall for Typhoon Megi or Haitang exceeds 350 mm over the 200-year return period. The risk map over the 200-year return period was generated in this study. The risk maps with various flood probability distributions would be recommended in the development of flood loss and risk assessment. However, the uncertainty in flood loss and risk assessment can lead to significant over-estimations or under-estimations [27]. The main uncertainties lie in the scarcity of data, the assessment model, and the human dimension e.g., the uncertainties within DEM accuracy, parameterization of the hydraulic model, house types, and field surveys [27,28]. The further study will address the impact of uncertainty in flood loss and risk assessment. In addition, the effects of the urban imperviousness [29], land development, and climate change can be quantified in hazard mitigation planning.

## 4. Conclusions

This study offers a GIS-based flood loss and risk assessment model to analyze the direct loss and risk impacts caused by floods in a residential urban area with small-sized residential buildings. Since the pattern of a residential area in the study area is dispersed, the water depths of a residential area are easier to identify using the high-resolution model. This study integrates high-resolution inundation simulation to generate fine-resolution flood hazard and risk maps. The high-resolution model provides detailed information of water depth in each pixel inside the residential area. The effect of the simulation resolution on flood loss was identified. The spatial resolution of simulation greatly affects the results of loss value. The assessment model shows different resolution estimations of a flood-inundated area and its corresponding loss. Moreover, vulnerability factors used in this study were prepared based on the importance of each of the vulnerability value factors in the study area. The AHP identified the weights of the vulnerability factors. The risk map is the result of combining flood hazard and vulnerability maps. Spatial flood risk maps show that the southwestern parts of the district near the river are most vulnerable to floods.

Future studies will consider flood risk under various return periods in order to identify the expected annual damage for each return period. Data collection of filled survey and social vulnerability factors, such as household income, and uncertainty analysis must also be added to develop a reliable loss and risk estimator.

**Author Contributions:** Z.A. contributed to model development, paper writing, and reporting of the results. H.-J.C. contributed to the study design, the model concept, data preparation, and the writing of the manuscript. Y.-L.K. and Y.-C.H. contributed to the improvement of the draft manuscript. H.K.W. contributed to the 3Di model. M.Z.A. contributed to paper processing. All authors have seen and approved the final version.

**Funding:** MOST 107-2119-M-006-024 and 107-2622-M-006-001-CC2 funded the APC.

**Acknowledgments:** The students from the International Master Program on Natural Hazards Mitigation and Management, NCKU, supported this work. The authors thank reviewers for valuable comments and suggestions.

**Conflicts of Interest:** The authors declare no conflict of interest.

## References

1. Teng, W.-H.; Hsu, M.-H.; Wu, C.-H.; Chen, A.S. Impact of Flood Disasters on Taiwan in the Last Quarter Century. *Nat. Hazards* **2006**, *37*, 191–207. [CrossRef]
2. Eves, C. The long-term impact of flooding on residential property values. *Prop. Manag.* **2002**, *20*, 214–227. [CrossRef]
3. Li, K.; Wu, S.; Dai, E.; Xu, Z. Flood loss analysis and quantitative risk assessment in China. *Nat. Hazards* **2012**, *63*, 737–760. [CrossRef]
4. Eleuterio, J.; Martinez, D.; Rozan, A. Developing a GIS tool to assess potential damage of future floods. *WIT Trans. Inf. Commun. Technol.* **2010**, *43*, 381–392.

5. Velasco, M.; Cabello, À.; Russo, B. Flood damage assessment in urban areas. Application to the Raval district of Barcelona using synthetic depth damage curves. *Urban Water J.* **2016**, *13*, 426–440. [CrossRef]
6. Penning-Rowsell, E.; Johnson, C.; Tunstall, S.; Tapsell, S.; Morris, J.; Chatterton, J.; Green, C. *The Benefits of Flood and Coastal Risk Management: A Manual of Assessment Techniques*; The Multi-Coloured Manual; Flood Hazard Research Centre, Middlesex University: London, UK, 2005.
7. Pedrono, M.; Locatelli, B.; Ezzine-de-Blas, D.; Pesche, D.; Morand, S.; Binot, A. *Impact of Climate Change on Ecosystem Services. Climate Change and Agriculture Worldwide*; Springer: Berlin/Heidelberg, Germany, 2016; pp. 251–261.
8. Naso, S.; Chen, A.S.; Aronica, G.T.; Djordjević, S. A novel approach to flood risk assessment: The Exposure-Vulnerability matrices. *E3S Web Conf.* **2016**, *7*, 08007. [CrossRef]
9. Shen, D.; Wang, J.; Cheng, X.; Rui, Y.; Ye, S. Integration of 2-D hydraulic model and high-resolution lidar-derived DEM for floodplain flow modeling. *Hydrol. Earth Syst. Sci.* **2015**, *19*, 3605–3616. [CrossRef]
10. Yang, S.-Y.; Chan, M.-H.; Chang, C.-H.; Chang, L.-F. The Damage Assessment of Flood Risk Transfer Effect on Surrounding Areas Arising from the Land Development in Tainan, Taiwan. *Water* **2018**, *10*, 473. [CrossRef]
11. Dahm, R.; Hsu, C.-T.; Lien, H.-C.; Chang, C.-H.; Prinsen, G. (Eds.) Next generation flood modelling using 3Di: A case study in Taiwan. In Proceedings of the DSD International Conference, Hongkong, China, 12–14 November 2014.
12. Stelling, G.S. Quadtree flood simulations with sub-grid digital elevation models. *Proc. Inst. Civ. Eng.* **2012**, *165*, 567–580.
13. Meesuk, V. *Point Cloud Data Fusion for Enhancing 2D Urban Flood Modelling*; CRC Press: Boca Raton, FL, USA, 2017.
14. Merz, B.; Hall, J.; Disse, M.; Schumann, A. Fluvial flood risk management in a changing world. *Nat. Hazards Earth Syst. Sci.* **2010**, *10*, 509–527. [CrossRef]
15. Birkmann, J. *Measuring Vulnerability to Natural Hazards: Towards Disaster Resilient Societies*; United Nations University: New York, NY, USA, 2006.
16. Roy, P.S. Flood risk assessment using multi-criteria analysis: A case study from Kopili River Basin, Assam, India. *Geomat. Nat. Hazards Risk.* **2018**, *9*, 79–93.
17. Karmakar, S.; Simonovic, S.P.; Peck, A.; Black, J. An Information System for Risk-Vulnerability Assessment to Flood. *J. Geogr. Inf. Syst.* **2010**, *2*, 129–146. [CrossRef]
18. Saaty, T.L. A scaling method for priorities in hierarchical structures. *J. Math. Psychol.* **1977**, *15*, 234–281. [CrossRef]
19. Goepel, K.D. A New AHP Excel Template with Multiple Inputs. 2013. Available online: http://bpmsg.com (accessed on 20 April 2018).
20. Nasiri, H.; Yusof, M.J.M.; Ali, T.A.M. An overview to flood vulnerability assessment methods. *Sustain. Resour. Manag.* **2016**, *2*, 331–336. [CrossRef]
21. Komolafe, A.; Herath, S.; Avtar, R. Sensitivity of flood damage estimation to spatial resolution. *J. Flood Risk Manag.* **2018**, *11*, S370–S381. [CrossRef]
22. Chen, A.S.; Hammond, M.J.; Djordjević, S.; Butler, D.; Khan, D.M.; Veerbeek, W. From hazard to impact: Flood damage assessment tools for mega cities. *Nat. Hazards* **2016**, *82*, 857–890. [CrossRef]
23. Bermúdez, M.; Zischg, A.P. Sensitivity of flood loss estimates to building representation and flow depth attribution methods in micro-scale flood modelling. *Nat. Hazards* **2018**, *92*, 1633–1648. [CrossRef]
24. Ward, P.J.; De Moel, H.; Aerts, J.C.J.H. How are flood risk estimates affected by the choice of return-periods? *Nat. Hazards Earth Syst. Sci.* **2011**, *11*, 3181–3195. [CrossRef]
25. Kobayashi, K.; Takara, K.; Sano, H.; Tsumori, H.; Sekii, K. A high-resolution large-scale flood hazard and economic risk model for the property loss insurance in Japan. *J. Flood Risk Manag.* **2016**, *9*, 136–153. [CrossRef]
26. Collet, L.; Beevers, L.; Stewart, M.D. Decision-Making and Flood Risk Uncertainty: Statistical Data Set Analysis for Flood Risk Assessment. *Water Resour. Res.* **2018**, *54*, 7291–7308. [CrossRef]
27. Wagenaar, D.J.; De Bruijn, K.M.; Bouwer, L.M.; De Moel, H. Uncertainty in flood damage estimates and its potential effect on investment decisions. *Nat. Hazards Earth Syst. Sci.* **2016**, *16*, 1–14. [CrossRef]

28. Winter, B.; Schneeberger, K.; Huttenlau, M.; Stötter, J. Sources of uncertainty in a probabilistic flood risk model. *Nat. Hazards* **2018**, *91*, 431–446. [CrossRef]
29. Lepeška, T. The impact of impervious surfaces on ecohydrology and health in urban ecosystems of Banská Bystrica (Slovakia). *Soil Water Res.* **2016**, *11*, 29–36. [CrossRef]

water

MDPI

Article

# The Development and Application of the Urban Flood Risk Assessment Model for Reflecting upon Urban Planning Elements

**Kiyong Park [1] and Man-Hyung Lee [2,*]**

1 Department of Disaster Prevention, Chungbuk National University, 1 Chungdae-ro, Seowon-gu, Cheongju, Chungbuk 28644, Korea; pky3489@chungbuk.ac.kr

2 Department of Urban Engineering, Chungbuk National University, 1 Chungdae-ro, Seowon-gu, Cheongju, Chungbuk 28644, Korea

* Correspondence: manlee@chungbuk.ac.kr; Tel.: +82-10-7623-2369

Received: 15 March 2019; Accepted: 25 April 2019; Published: 1 May 2019

**Abstract:** As a city develops and expands, it is likely confronted with a variety of environmental problems. Although the impact of climate change on people has continuously increased in the past, great numbers of natural disasters in urban areas have become varied in terms of form. Among these urban disasters, urban flooding is the most frequent type, and this study focuses on urban flooding. In cities, the population and major facilities are concentrated, and to examine flooding issues in these urban areas, different levels of flooding risk are classified on 100 m × 100 m geographic grids to maximize the spatial efficiency during the flooding events and to minimize the following flooding damage. In this analysis, vulnerability and exposure tests are adopted to analyze urban flooding risks. The first method is based on land-use planning, and the building-to-land ratio. Using fuzzy approaches, the tests focus on risks. However, the latter method using the HEC-Ras model examines factors such as topology and precipitation volume. By mapping the classification of land-use and flooding, the risk of urban flooding is evaluated by grade-scales: green, yellow, orange, and red zones. There are two key findings and theoretical contributions of this study. First, the areas with a high flood risk are mainly restricted to central commercial areas where the main urban functions are concentrated. Additionally, the development density and urbanization are relatively high in these areas, in addition to the old center of urban areas. In the case of Changwon City, Euichang-gu and Seongsan-gu have increased the flood risk because of the high property value of commercial areas and high building density in these regions. Thus, land-use planning of these districts should be designed to reflect upon the different levels of flood risks, in addition to the preparation of anti-disaster facilities to mitigate flood damages in high flood risk areas. Urban flood risk analysis for individual land use districts would facilitate urban planners and managers to prioritize the areas with a high flood risk and to prepare responding preventive measures for more efficient flood management.

**Keywords:** climate change; urban flood risk; flood damage; urban disaster; land use

## 1. Introduction

The degree and scale of flood hazards have massively increased with the changing climate in the last decades. The larger-scale flash floods than in the past have brought fast-moving and rapid-rising water with force, resulting in tremendous life and property losses, as well as social disruption worldwide [1].

Floods are natural processes in river systems [2,3]. However, humans have occupied and urbanized floodplains for their urbanization attractiveness due to their planar morphology and water availability [4,5]. The latter has translated into the growth of flood risk zones for human settlements and infrastructure due to a greater concentration of people and structures [6,7]. Recent urban growth has not taken the space that rivers require to temporarily store flows during floods into consideration [4]. Global efforts have focused more on implementing flood control infrastructure, such as dikes, dams, and channelization, but despite these efforts, modern cities still remain vulnerable to flood risk [8].

Furthermore, the potential for flood casualties and damages is also increasing in many regions due to the social and economic development, which implies pressure on land-use, e.g., through urbanization. Flood hazard is expected to increase in frequency and severity, through the impacts of global change on climate, resulting in severe weather in the form of heavy rains and river discharge conditions [9].

More cities are becoming hotspots for risk and disaster [10], mainly as a result of rapid urbanization, population growth, and the impacts of climate change [11].

Throughout the world, cities have been affected by the increasing impacts of floods. In the period 1998–2008, more than 2900 events were registered [12]. Recent estimates indicate that urban zones exposed to flooding will increase 2.7 times by the year 2030 [13]. Globally, the previewed scenario demonstrates an increase in the frequency and magnitude of floods due to the changes in precipitation patterns resulting from climate change and accelerated urban expansion [14–17]. It is estimated that by 2050, 70% of the world's population will be concentrated in urban areas [13,15,18].

Countermeasures to urban flooding should be considered in long-term perspectives because the impacts of climate change are unpredictable and complex [19].

The development of appropriate flood risk management strategies for flooding should be considered in long-term perspectives (e.g., expected future rainfall amounts, although climate change impacts can be unpredictable and complex) and should focus on increasing an area's resilience to flooding. Recently, as rainfall has become more concentrated over short periods of time, substantial amounts of damage due to pluvial flooding have occurred in urban areas. This includes damage of social infrastructures, as well as losses of human life and properties. Conventionally, storm water management has been focused on drainage systems via underground pipes. However, these conventional approaches (structural measures) have had problems in many cases because they were designed based on historical events. This makes it difficult to deal with extreme rainfall events that exceed the designated capacity. The frequency of extreme rainfall events is expected to increase with projected climate change, which may cause conventional stormwater management systems to be exceeded [20].

While information on the distribution of flood hazard at a national to global scale is extremely valuable, knowledge about how it is distributed with reference to the population and built assets is critical to water resource managers, city planners, and policy-makers [21].

In the past, natural disasters were accepted as an unavoidable calamity, and the emphasis was placed on how to cope with the aftermath of the disaster, rather than how to prevent it. However, the prevention measure shall shift its focus to preemptive prevention from a holistic point of view considering the large scale, unexpectedness, and complexity of the disaster. This study suggests the development and application of the urban flood risk assessment model to reflect upon urban planning elements (land use, building characteristics). In this respect, measuring in terms of urban planning is the most effective preemptive measure to reduce damage from natural disasters, and it focuses on the idea that the development shall be inhibited in the area that is vulnerable to flood damage and carries a high risk from the stage of land utilization planning in order to minimize the damage. In addition, an urban utilization plan may be an important policy measure in achieving the goal of keeping the city safe by restricting development in areas vulnerable to natural disaster. It is important to note that the impact of urban land use on disaster damages may escalate when unplanned and thoughtless developments prevail without the construction of appropriate infrastructure facilities and that this may become a reality while damages are currently increasing in urban areas due to reckless land use. Reducing the area for urban land use would naturally reduce the resilience of disaster damages. However, it would be difficult to expect such a reduction in the area for urban land use as there has not been any reduction recently. Therefore, the purpose of this study is to analyze the flood risk by concentrating on land use and building characteristics reflecting on urban planning elements. The flood risk is classified to maximize the spatial efficiency during the flooding events and to minimize the following flooding damage. The paper concludes by stressing the necessity of land use planning measures to prevent flooding. It is possible to classify flood risk areas according to the use of districts and administrative districts, evaluate priorities, and enable efficient management by selecting areas with a high flood risk.

## 2. Methods

In Sections 2.1 and 2.2, the theoretical perspectives are presented. Following this, Section 2.3 explores the relationship between urban space and flooding. In Section 2.4, the overall process of this study is described. Furthermore, in Section 2.5, the evaluation item of the analysis is presented. In Section 2.6, the fuzzy classification is introduced. Finally, in Section 2.7, the background of the selection of the analysis area is given.

### 2.1. Vulnerability and Exposure of Urban Flood

The risk of disaster is discontinuous and local. The only way to reduce the risk is to reduce the vulnerability within the system. The risk that places stress on the system is purely natural and is caused by variations outside of the system, and there is nothing that people can do to reduce the risk itself. However, human beings can only reduce the vulnerability to natural disasters by changing social systems or social infrastructure.

The concept of vulnerability defined by the IPCC is based on the view that combines vulnerability as the result of external stress and vulnerability as an internal state of the system. Vulnerability is the state of being easily affected by adverse impacts of climate change, including climate variations and extreme events, or the degree of inability to cope with it, and is the function of the characteristics, scale, and speed of climate variations to which a system is exposed and the sensitivity and susceptibility of the system [22,23]. In this respect, vulnerability should be regarded as potential exposure to damage rather than an estimation of damage due to a sudden change in climate or stress based on probability.

In the 8$^{th}$ IHP Strategic Plan of UNESCO, the specific concept of vulnerability was defined as in Figure 1, based on the disaster characteristics-damage relationship, by focusing on the idea that there is a difference in the scale of damage due to vulnerability, even for natural disasters of the same intensity. As explained above, the conceptual definition of vulnerability may be slightly different. However, it can be summarized that when the impact of natural disasters is great, the vulnerability of a system is considered high if it has a small capacity to cope with it. On the contrary, the capacity to cope with disaster can be high, even when the impact of natural disasters is high. This system can have the opportunity for development while coping with disasters appropriately. If the system has a small capacity to cope and the natural disaster has a small impact, the system may still have residual risks. If the impact is small and the system has a great capacity to cope, the system may promote sustainable development.

Degree of exposure has been defined as the person, property, system, or other components in dangerous areas affected by a potential loss and can be measured by the number of people and assets in the area. To quantitatively estimate the risk involved with the hazard, the exposure can be combined with specific vulnerabilities of the elements exposed to a specific risk.

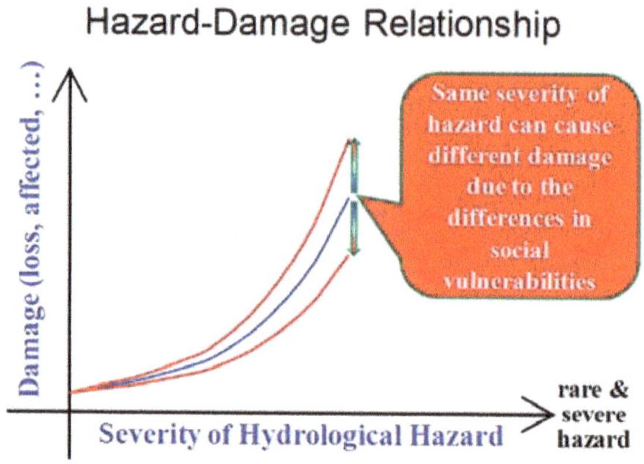


**Figure 1.** Relationship between the damage caused by hydrological extremes and the severity of the event [24].

### 2.2. Risk of Urban Flood

The risk can be defined as the possibility that the loss will be greater than what was generally expected. In other words, risk is a concept of probability based on possibility, but it can be re-defined as the difference between expectation and reality considering the amount of loss [25].

According to ISO 31000, risk is defined as the "effect of uncertainty toward an object" or "the combination of the probability that an event would occur and the outcome of the event." The probability that an event will occur is related to the source of the disaster and its properties, and the outcome is related to vulnerability, which influences the scale of damage and the capacity to reduce damage.

There are still many cases where the meaning of risk is not clearly defined and is used confusedly to mean the degree of risk, hazardousness, or vulnerability. In this study, the components of risk considered are riskiness, vulnerability, and exposure, and also include the ability to adapt to risks.

### 2.3. Relationship between Urban Space and Flooding

Urban spaces carry a high risk of massive and complex human and property damages as they are quite vulnerable to disasters due to the concentration and densification of the population and facilities, the increased interdependence of various urban facilities and activities, the development of lowlands and slopes, and the development of underground space [26].

According to the fifth report of the IPCC, the artificial activity of human beings accounts for an absolute ratio of 95% of the cause of climate change. Paradoxically, however, human beings are also the ones that are most affected by diversified and escalated damages from various disasters caused by climate change. The city is a space where such human activities are concentrated, and the city, as well as the environment that surrounds people, influence and are influenced by climate change. Recent studies on disasters show that the impact of a natural disaster is caused by the interaction between natural phenomena and people [27]. Eventually, the spaces that explain this phenomenon gather and form a city.

Urbanization and climate change have direct and indirect impacts on disasters or cause such disasters within urban spaces. Urbanization and climate change, which have great relevance across all areas of the urban space, are considered to be important paradigms related to natural disasters. If urbanization was the global phenomenon that drew the attention of the whole world in the 1900s, the international issue of the 2000s is climate change. The most common opinion among experts is that the influence of climate change will appear more prominently in society along with the acceleration of urbanization [28]. They argue that climate change is increasing natural disaster damage around the world, and the trend of urbanization is escalating damages [22].

The urbanization rate around the world expanded from 23.8% in 1950 to 50% in 2010. Urbanization is a population-concentrating phenomenon that appears after the concentration of various facilities and functions, and it is accompanied by various environmental problems [29]. The KOSIS (Korean Statistical Information Service) data also predicted that the urbanization rate will reach 60% or over as of 2030. Although the increase rate of urbanization has decreased after exceeding 50% in 2010, the rate of urbanization itself is expected to increase continuously. Assuming that the current environment remains the same, it is clear that the damage resulting from natural disasters will increase [30].

Urbanization is concentrating the limited urban space with artificial factors such as population, industries, and facilities, and increasing urban spaces vulnerable to natural disasters through indiscreet land use. In particular, unpredictable natural disasters due to climate change not only increase the risk even further in urban spaces with high vulnerability and urban spaces with less recovering capabilities but also form a vicious cycle of causing damage from natural disasters.

### 2.4. Methods of Risk Analysis

The basic ideas and concepts are shown above. The vulnerability analysis was carried out based on urban space characteristics and building characteristics, including non-structural characteristics, by setting 100 m × 100 m geographic grids as the evaluation units alongside urban flooding analysis based on environmental factors such as topographical characteristics and rainfall characteristics. The urban flood risk was assessed and analyzed based on these results and using the risk mapping technique. Concepts and processes important to their construction are described herein and in Figure 2.

1) Vulnerability analysis: Vulnerability is the result of external stress, which implies an internal condition of urban areas. It is the degree of socio-economic damage to the urban space (land use) and building characteristics (land price, floor area ratio, underground area, decline of building, material of building) during flooding.
2) Exposure (Hazard) analysis: An empirical analysis is of the flood depth, and flooding area was conducted considering the environmental factors, such as rainfall and topography. It is frequently referred to as an "inundation trace map", which is the result of an urban flooding map.
3) Flood risk analysis: This established the flood risk evaluation model analysis system by reflecting on the land use and building characteristics which are included in the non-structural measures to perform evaluation and analysis. Flood risk was calculated by deriving a value from each grid and overlapping it with the map by considering the vulnerability data, which classified the risk in each grid unit, and the analysis data for flooded areas, such as disaster characteristics, were employed as a function.
4) Urban flood risk assessment: Areas with high and low risks were determined by objective and scientific methods based on the urban flood risk assessment. Finally, the intent was to establish a long-term measure by setting the flood risk area based on the results analyzed by these objective and scientific methods.

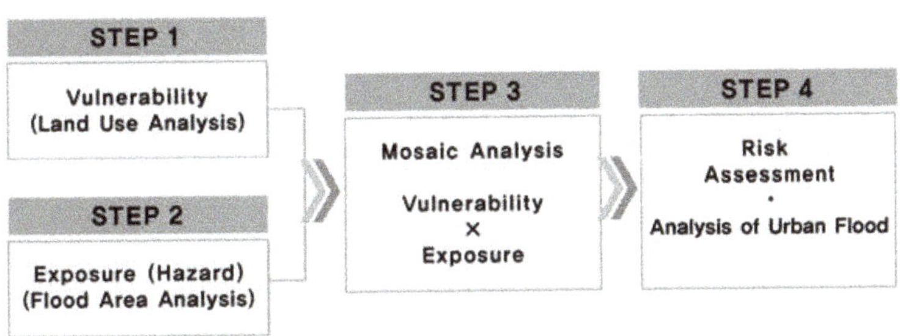


**Figure 2.** Analysis framework of urban flood risk.

### 2.5. Evaluation Item

Objective indicators and values were calculated to analyze the risk of flooding. Each indicator was selected based on the following standards.

The land price for each use district provided by the Korea Appraisal Board is considered an important indicator of flood damage because property damage represents the major loss in the case of a flood [31]. As the Ministry of Construction and Transportation established long-term water resource use planning in 2001, local characteristics of water resources were carefully examined to identify investment priority in watershed planning and development. As a result, Potential Flood Damage (PFD) was estimated, for which the property value plays an important role [32]. In the research of Han [33], the correlation between various indicators of flood vulnerability and the damage costs for flooding from 1971 and 2000 in different watersheds was analyzed. As a result, the property density had a significant correlation with the damage cost of flooding, which was a result attributed to the estimation method of flood damage cost largely reflecting the land price [33].

The underground area index is directly related to the flood reference system to prevent building submergence within the flood-water disaster prevention criterion for buildings [34]. In a study of Jang [35], the vulnerability of buildings due to flooding was found to be closely related to the presence and size of underground space. The top 10 items with the greatest damage were located in places where there were significant portions of underground space, while the bottom 10 items had no underground space. Therefore, the size of underground space was important for determining the vulnerability of buildings. Additionally, it is considered that rainwater adversely affects the building structures by moisture, warping, cracks, and corrosion when rainwater fills the underground space or penetrates through the wall [35].

The floor area ratio is a concept of disaster that collectively refers to the case where urban spaces lose their function and damages are multiplied due to a high density. Because of the high utilization of urban space, complex factors such as underground space and humans instigate unpredictable large-scaled disasters. In the United States, the Community Rating System (CRS), which was established in 1990 to operate the flood insurance system, has been employed to evaluate the building height issued by the Federal Emergency Management Agency (FEMA) for 19 items. Since the damage is expected to vary depending on the area and density of the building, the indicators are selected by considering urban functional damage [31].

A decline of building shows that the incidence of accidents is frequent as the durability reaches its limit. Moreover, the development-centered paradigm of urban planning has escalated an event to a series of failures, resulting in a disaster. Therefore, urban disasters have a high potential to result in a collective paralysis of urban functions [32].

The material of buildings is a robustness-related index. Unanwa et al. (2000) stated that property values of the building and exterior wall types were set. In the case of building interiors, it has been shown that many damages occur during flooding due to the heavy use of wood [36].

### 2.6. Fuzzy Classification

Fuzzy set theory has been developed and extensively applied since 1965 [37]. It was designed to supplement the interpretation of linguistic or measured uncertainties for real-world random phenomena. These uncertainties could originate with non-statistical characteristics in nature that refer to the absence of sharp boundaries in information. However, the main source of uncertainties involved in a large-scale complex decision-making process may be properly described via fuzzy membership functions [38].

Fuzzy classification is, alongside neural networks [39] and probabilistic approaches [40], a very powerful soft classifier. As an expert system for classification [41], it takes into account uncertainty in sensor measurements, parameter variations due to limited sensor calibration, vague (linguistic) class descriptions, and class mixtures due to a limited resolution. Fuzzy classification consists of an n-dimensional tuple of membership degrees, which describes the degree of class assignment μ of the considered object obj to the n considered classes.

$$f_{class,obj} = \left[\mu_{class_1}(obj), \mu_{class\_2}(obj), \ldots \mu_{class\_n}(obj)\right] \quad (1)$$

Crisp classification would only provide information on which membership degree is the highest, whereas this tuple contains all information about the overall reliability, stability, and class mixture. Fuzzy classification requires a complete fuzzy system, consisting of the fuzzification of input variables, resulting in fuzzy sets, fuzzy logic combinations of these fuzzy sets, and defuzzification of the fuzzy classification result to get the common crisp classification for map production. Fuzzy logic is a multi-valued logic quantifying uncertain statements. The basic idea is to replace the two Boolean logical statements "true" and "false" by the continuous range from 0 to 1, where 0 means "false" and one means "true", and all values between 0 and 1 represent a transition between true and false. Avoiding arbitrary sharp thresholds, fuzzy logic is able to approximate real-world complexity much better than the simplifying Boolean systems do. Fuzzy logic can model imprecise human thinking and can represent linguistic rules. Hence, fuzzy classification systems are well-suited to handle most sources of vagueness in remote sensing information extraction. The mentioned parameter and model uncertainties are considered by fuzzy sets, which are defined by membership functions. Fuzzy systems consist of three main steps, including fuzzification and the combination of fuzzy sets [42].

In the flood damage risk classification analysis, it is very difficult to quantify and compare the flood damage in commercial and residential areas when the same area is submerged. In other words, it is ambiguous to express the level of flood damage as numerical values. Therefore, to overcome the linguistic ambiguity for decision-making in previous studies and to analyze complex relationships between different indicators and indices, the fuzzy logic method is adopted to perform a more objective analysis of flood risk by deriving quantitative and accurate indicators.

### 2.7. Analysis Area

The distribution of flood damage in Korea analyzed by the Korea Ministry of Land, Infrastructure, and Transport is shown in Figure 3 [37], and Changwon City is selected as the target area due to its advantages of data construction because it includes areas with both high and low flood damage [31].

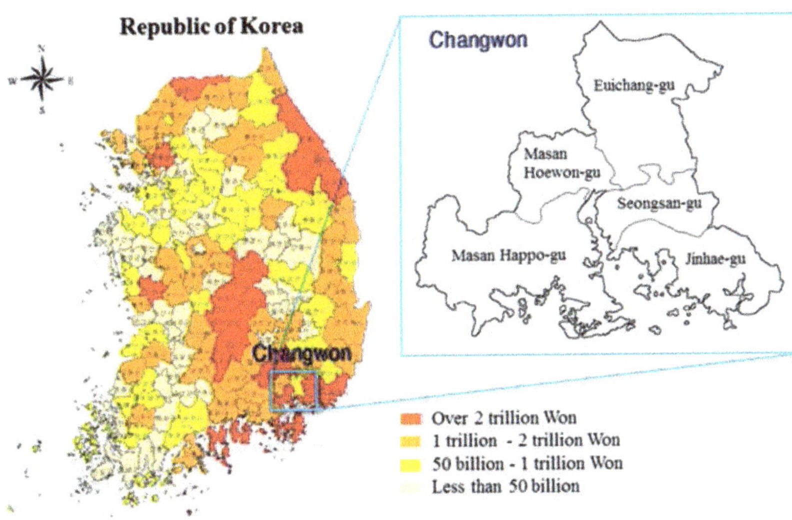


**Figure 3.** Distribution of national flood damage in Korea [43].

Changwon City has become one of the first successful administrative integration models in the nation as it integrated Changwon, Masan, and Jinhae cities in July 2010, and it is becoming the first growth base of the Southeastern Greater Economic Zone in Korea. The administrative district consists of the five districts of Masan Happo-gu, Masan Hoewon-gu, Seongsan-gu, Euichang-gu, and Jinhae-gu, 351 legal smaller districts, and 62 administrative smaller districts. The administrative area of Changwon City is 746.58 $km^2$ [44].

As a result of examining the land use situation from 1975 to 2007, using the land cover map of Changwon City, the urbanization rate greatly increased from 3.95% in the 1990s to 13.61% in recent years. In addition, a countermeasure to flood damage in Changwon city is needed because the urbanization rate is planned to increase from 14.13% in 2020 to 15.72% in 2025, according to the step-by-step development plan of Changwon City Basic Plan 2025 [44]. In response, the risk according to flood damage was analyzed in Changwon City.

## 3. Results and Discussion

### 3.1. Analysis of Vulnerability

To select the indicators capable of assessing the flood vulnerability to climate change, indicators related to land use and building in the non-structural aspect that were proposed in previous research were selected, and the final indicators were selected through content validity analysis. To calculate the quotient, the fuzzy methodology was used to derive the position function, and the fuzzy inference rules were established and implemented to analyze the vulnerability of the districts of Changwon, as seen in Figure 4. As all the spaces, buildings, and facilities in the city are basically the subject of disaster prevention, the characteristics of the areas were analyzed in 100 m × 100 m geographic grids, which maintains the livelihood of the citizens and the urban function.

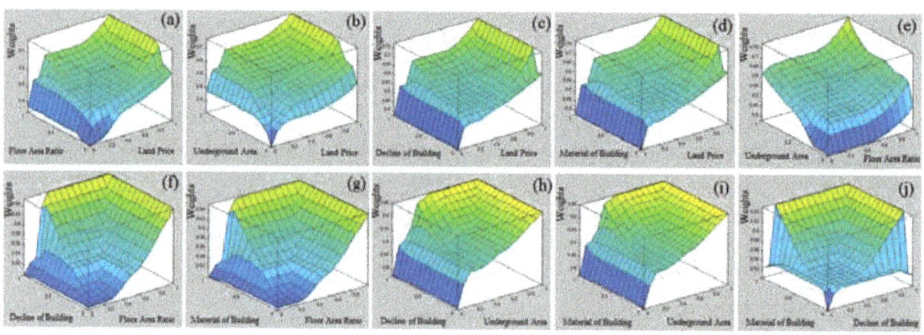

**Figure 4.** Flood damage scheme: (**a**) land price—floor area ratio, (**b**) land price—underground area, (**c**) land price—decline of building, (**d**) land price—material of building, (**e**) floor area ratio—underground area, (**f**) floor area ratio—decline of building, (**g**) floor area ratio—material of building, (**h**) underground area—decline of building, (**i**) underground area—material of building, and (**j**) decline of building—material of building.

There were a total of 177,193 cases of the officially assessed land prices for each lot in Changwon city, and the data were constructed using the individual officially assessed land prices provided by the Korea Appraisal Board. There were a total of 66,269 cases of the floor area ratio, 15,021 cases of underground area, 97,346 construction declines of buildings, and 69,932 cases of building materials according to the building register. The map of Changwon City was divided into 77,737 cells of 100 m × 100 m geographic grids, and maximum and minimum values, as well as the average value, were standardized for the officially assessed land price, floor area ratio, underground area, decline of building, and building material indices, and fuzzy analysis was conducted to evaluate the vulnerability. As a result of the fuzzy analysis, the index with the highest vulnerability to flooding was land price, followed by underground area, floor area ratio, decline of building, and material of building. The implication of the result is that the land price and the underground space are direct indicators of property damage and physical damage in the case of flooding, whereas the floor area ratio is directly related to the only corresponding floor and indirectly related to other floors for its inconvenience in urban functional aspects. Furthermore, it is shown that decline of building and material of building are robustness-related indexes, which is less important than other indicators.

A higher fuzzy score indicates a higher degree of flood risk (greater flood damage), and a lower fuzzy score indicates a lower degree of flood risk (no damage). The fuzzy values for 100 m × 100 m geographic grids were derived from fuzzy analysis through standardized values of each indicator for officially assessed land price, floor area ratio, underground area, decline of building, and building materials, listed in Table 1.

**Table 1.** Fuzzy values for 100 m × 100 m geographic grids.

| | Standard Values | | | | | Fuzzy |
|---|---|---|---|---|---|---|
| | Land Price | Floor Area Ratio | Underground Area | Decline of Building | Material of Building | |
| 1 | 0.5408 | 0.7624 | 0.0529 | 0.0356 | 0.0272 | 0.6150 |
| 2 | 0.6134 | 0.1774 | 0.0107 | 0.0910 | 0.0243 | 0.5552 |
| 3 | 0.6134 | 0.5305 | 0.0055 | 0.0742 | 0.0063 | 0.5488 |
| 4 | 0.6396 | 0.5612 | 0.0049 | 0.0425 | 0.0066 | 0.5475 |
| 5 | 0.5185 | 0.5996 | 0.0107 | 0.0430 | 0.0082 | 0.5454 |
| 6 | 0.1084 | 0.0001 | 0.0001 | 0.0001 | 0.0001 | 0.0950 |
| 7 | 0.1092 | 0.0925 | 0.0001 | 0.0356 | 0.0142 | 0.2200 |
| 8 | 0.0900 | 0.0001 | 0.0001 | 0.0341 | 0.0079 | 0.0950 |
| 9 | 0.1065 | 0.1136 | 0.0272 | 0.0346 | 0.0126 | 0.2272 |
| 10 | 0.1028 | 0.1067 | 0.0001 | 0.0338 | 0.0081 | 0.2200 |
| 11 | 0.1028 | 0.1335 | 0.0001 | 0.0326 | 0.0078 | 0.2500 |
| 77731 | 0.1779 | 0.0001 | 0.0001 | 0.0001 | 0.0001 | 0.2200 |
| 77732 | 0.2473 | 0.0967 | 0.0014 | 0.0059 | 0.0028 | 0.2377 |
| 77733 | 0.173 | 0.2219 | 0.0128 | 0.0089 | 0.0059 | 0.3388 |
| 77734 | 0.173 | 0.0001 | 0.0001 | 0.0001 | 0.0001 | 0.2200 |
| 77735 | 0.0979 | 0.0924 | 0.0023 | 0.0831 | 0.0028 | 0.2200 |
| 77736 | 0.0124 | 0.0473 | - | 0.0593 | 0.0042 | 0.0950 |
| 77737 | 0.0376 | - | - | - | - | 0.0950 |

Based on the fuzzy analysis value of each indicator, 100 m × 100 m geographic grids were prioritized for flood damages. When the same area was flooded, the area with the most flood damage in a social and economic sense, which is the area with high vulnerability, was analyzed as the area with the highest development density in Changwon centering on the central commercial area. In addition, the areas that played key functions in the city in the past as the old town center, although currently declining, showed high vulnerability.

Once the vulnerability grade for land use has been determined based on standardization of the fuzzy score, it is possible to classify based on the level of risk. The standard color was determined by designating the area with the highest vulnerability as the red zone, the area with high vulnerability as the orange zone, the area with intermediate vulnerability as the yellow zone, and the area with low vulnerability as the green zone. If the resulting values are applied to the map of Changwon, it is possible to derive the land use classification map, as shown in Figure 5.

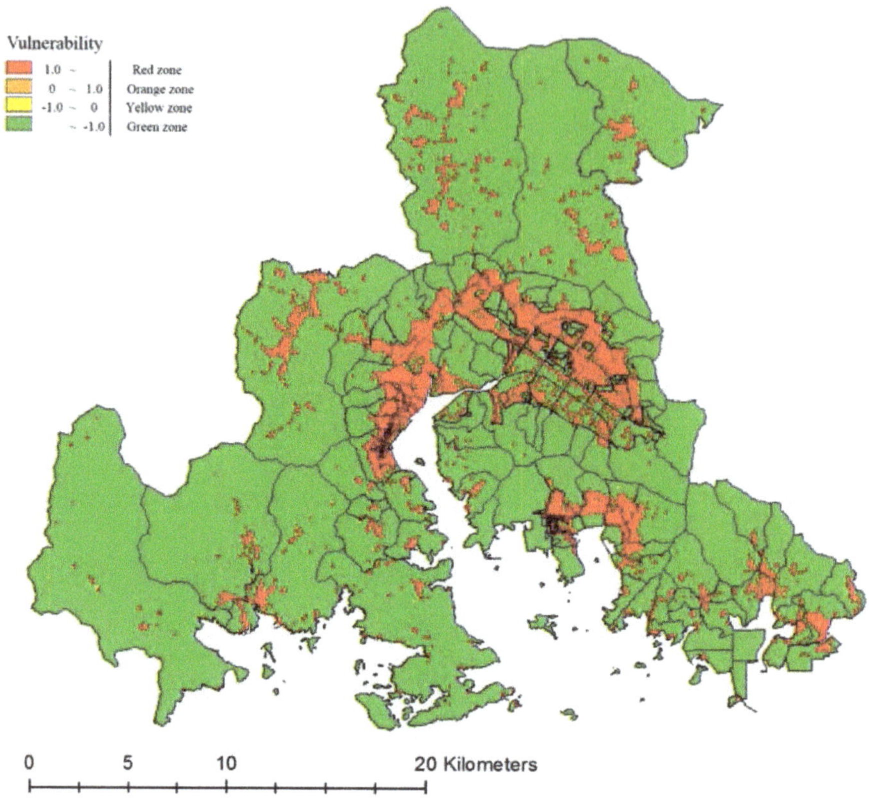


**Figure 5.** Urban flood vulnerability analysis.

### 3.2. *Analysis of Exposure*

The analysis using HEC-Ras is a typical flood analysis model that conducts a simulation by constructing the river crossing data, the center line of the river, the river bank line, and and using the flood amount and the flood level as the boundary conditions. This analysis process is shown in Figure 6.

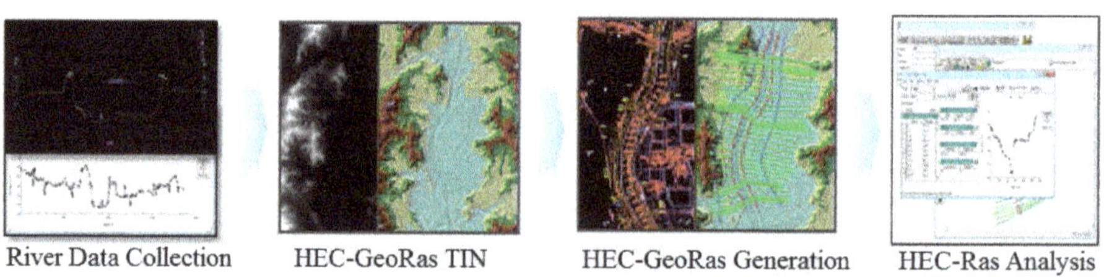


**Figure 6.** Flood depth and flooded area analysis using HEC-Ras.

The urban flood map prepared by using the one-dimensional model HEC-Ras based on the basic environmental factors such as rainfall data and topographical data developed by the Ministry of Interior and Safety was used in this study [45]. The urban flood map is shown in Figure 7 below. The urban flooding area evaluation confirmed that the flood depth was high in the northern area adjacent to the river and reservoir and also high in the areas adjacent to Changwon City Hall, which play key roles in Changwon City and are also considered as the town center.

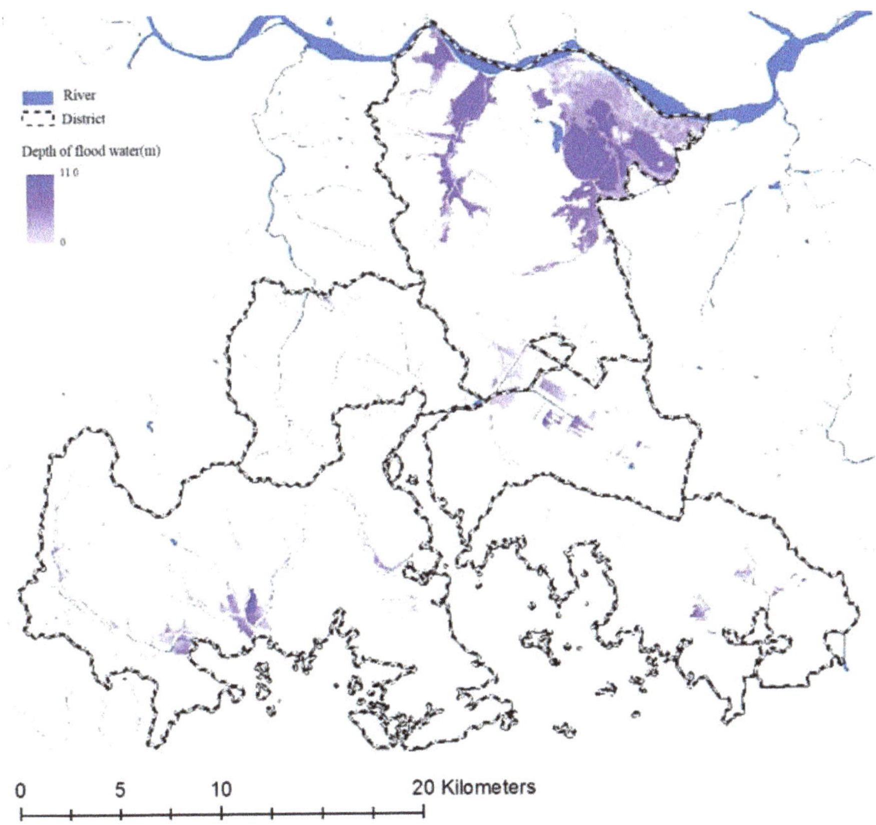


**Figure 7.** Urban flood exposure analysis.

### 3.3. Analysis of Risk

Figure 8 displays the graded urban flood risk by overlapping the non-structural factors, which are the criteria for socioeconomic damages, derived based on the study by Kron [46] and Brooks et al. [47], with the flood simulation analysis map. This map analyzed the flood depth based on the land use classification map, formed of vulnerability analysis data and data on disaster characteristics (rainfall, topography), including the degree of exposure.

The analysis of urban flood risk confirmed that the red zone was mainly found in the central commercial area, where various infrastructures, including public institutions that play key functions in the city, are located. This appears to be related to the development considering the economy, convenience, and efficiency of the city. The orange zone shows a fan-shaped distribution centering on the red zone, which also demonstrates the tendency of branching out from major functions and socio-economic parts of the city, and by use area, they were classified into distribution commercial areas, including general commercial areas and quasi-residential areas. The distribution of the yellow zone centered on the residential area confirms that it is distributed in the area that was formed a long time ago. It is considered that this is closely related to the aging of buildings. The green zone was mainly distributed in green areas where the environmental values or the environment rather than the value of development land use need to be preserved.

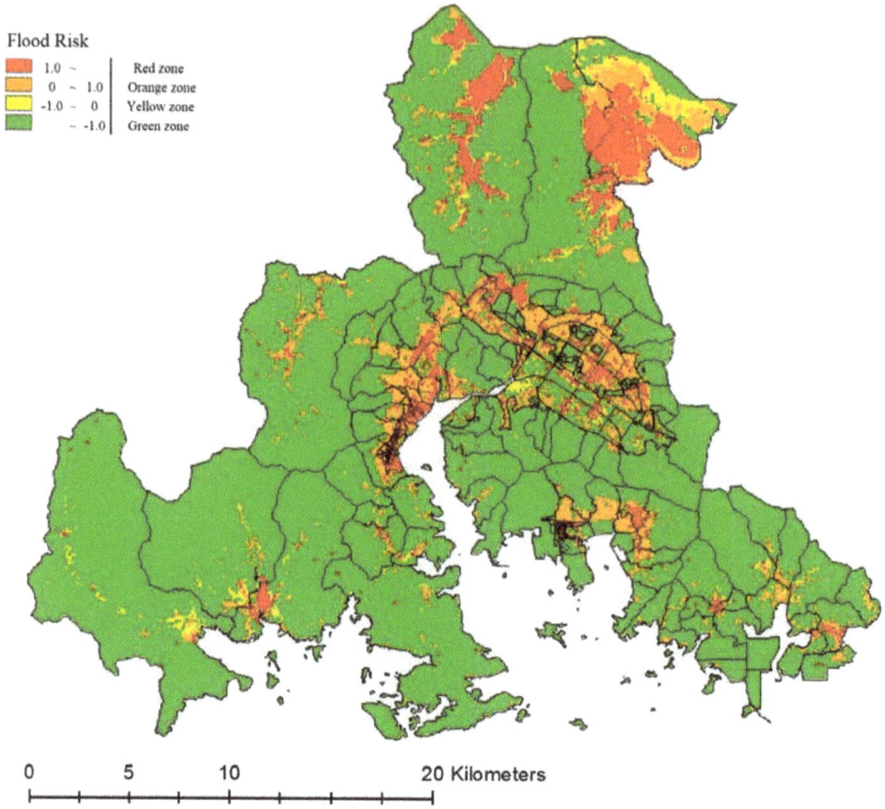


**Figure 8.** Urban flood risk analysis.

### 3.4. Overall Analysis Results

The results of the risk analysis by district are shown in Table 2. The area with the highest urban flood risk and the highest percentage of red zone in urban flood risk was Euichang-gu (13.07%), followed by Seongsan-gu (4.07%), Masan Hoewon-gu (2.68%), Masan Happo-gu (1.87%), and Jinhae-gu (1.78%), in respective order. The orange zone was also distributed more frequently in the window (14.78%) and in Seongsan-gu (12.37%) than in other areas. The distribution of the yellow zone was most prominent in Euichang-gu (3.96%), followed by Seongsan-gu (2.10%) and Masan Happo-gu (1.39%), respectively. The green zone, which is the safe area with the lowest urban flood risk, was most distributed in Masan Happo-gu (92.21%), Jinhae-gu (86.99%), and Masan Hoewon-gu (85.89%).

**Table 2.** Risk for each administrative district.

| | Area ($m^2$) | Green Zone | | Yellow Zone | | Orange Zone | | Red Zone | |
|---|---|---|---|---|---|---|---|---|---|
| | | Cells of Grid Units | Ratio (%) | Cells of Grid Units | Ratio (%) | Cells of Grid Units | Ratio (%) | Cells of Grid Units | Ratio (%) |
| Masan Happo-gu | 239,630,000 | 22096 | 92.21 | 332 | 1.39 | 1087 | 4.54 | 448 | 1.87 |
| Masan Hwewon-gu | 90,840,000 | 7802 | 85.89 | 31 | 0.34 | 1008 | 11.1 | 243 | 2.68 |
| Seongsan-gu | 82,080,000 | 6687 | 81.47 | 172 | 2.1 | 1015 | 12.37 | 334 | 4.07 |
| Euchang-gu | 211,320,000 | 14410 | 68.19 | 836 | 3.96 | 3124 | 14.78 | 2762 | 13.07 |
| Jinhae-gu | 120,410,000 | 10,475 | 86.99 | 81 | 0.67 | 1271 | 10.56 | 214 | 1.78 |

Masan Happo-gu has been active as the general commercial area since being developed a long time ago centering on Masan Port and served as the transportation hub with railroads in the past, although it is now an abandoned railroad site. The neighborhood commercial area and semi-residential area were formed around the district, and the urban flood risk in these areas appears to be high. The flood risk in the industrial areas is relatively low as they were recently formed sporadically in mountainous areas. Considering the results in terms of building characteristics, a high fuzzy value for construction year across the whole use area indicates that the city was formed long ago and that the aging of buildings is becoming serious. Additionally, the basement is distributed more in the industrial area than other use areas.

In Masan Hoewon-gu, a general commercial area is mainly formed around Masan Station and its neighborhood, with commercial areas, semi-residential areas, and general residential areas around it. The industrial area was extensively formed in the southeastern area in the past and the flood risk is more or less high as the vulnerability of the building materials is high. According to building characteristics, the construction year was higher than other indicators in the same way as Masan Happo-gu, indicating that the buildings are aged. In particular, the officially assessed land price of the neighboring commercial area is much higher than other indicators, and this suggested that there is a great demand for daily necessities and services from the residents of the neighborhood residential area and the area appeared to have been specialized for that purpose.

Seongsan-gu is a comparatively new city area, where the commercial area and residential area have been recently developed, centering on public institutions including the city hall, and the fuzzy value for the materials is low as the buildings are not aged and many of them were constructed recently. The officially assessed land price and floor area ratio are high throughout the use district, particularly in the general commercial area, central commercial area, and semi-residential area, indicating that it is an area with high asset value. The fuzzy value is high, particularly as the highly dense central commercial area is activated, and also as many semi-residential areas incorporate commercial facilities among the residential areas. The reason why the fuzzy value is low in the green area is that it has a low asset value and it is quite unlikely that development of the area would occur in this region.

Euichang-gu is located near Seongsan-gu. Like Seongsan-gu, the commercial areas and residential areas have recently been developed around public institutions, including the provincial government buildings, and the fuzzy value of the city's land price is considerably higher than other indicators, indicating that the asset value of the region is high. The green area demonstrated the lowest fuzzy value like other areas, indicating the lowest vulnerability.

Jinhae-gu has a low asset value and the fuzzy value of the construction year was high throughout the whole use district, indicating that the city has existed for a long time and the buildings are aged compared with other areas. In the Jinhae area, semi-residential areas and general commercial areas showed more or less a higher vulnerability, and fuzzy values of industrial areas and green zones were low.

By use district, the ratio of red zones of commercial areas, including central commercial areas and general commercial areas, was higher, as shown in Table 3, and the risk declines in the order of residential areas > industrial areas > green areas.

These results imply that the precautionary measures for flooded areas should primarily focus on the central commercial areas among the use district of the urban areas. In addition, the green zones have the least risk, as shown in the analysis results, and the potential to reduce or minimize the flood risk and may demonstrate more flexibility than the structural measures in various aspects considering the uncertainty of climate change. Moreover, the green zones are more effective than aged structural systems (sewage, storage facilities) in the long term, and not only reduce the rainwater runoff rate, but also reduce water pollution, which makes them one of the most useful resources, while the structural (sewer) system generates pollution and increases water pollution. Therefore, the results confirmed the fact that the flood risk could be reduced by arranging green zones appropriately.

**Table 3.** Risk for each use district.

| Use District | | Area ($m^2$) | Green Zone | | Yellow Zone | | Orange Zone | | Red Zone | |
|---|---|---|---|---|---|---|---|---|---|---|
| | | | Cells of Grid Units | Ratio (%) | Cells of Grid Units | Ratio (%) | Cells of Grid Units | Ratio (%) | Cells of Grid Units | Ratio (%) |
| Residential Area | Private Residential Area | 11,300,000 | 249 | 22.04 | 18 | 1.59 | 767 | 67.88 | 96 | 8.5 |
| | General Residential Area | 44,000,000 | 1374 | 31.23 | 63 | 1.43 | 2265 | 51.48 | 698 | 15.86 |
| | Semi-residential Area | 2,080,000 | 43 | 20.67 | 3 | 1.44 | 106 | 50.96 | 56 | 26.92 |
| Commercial Area | Central Commercial Area | 1,230,000 | 37 | 30.08 | 0 | 0 | 44 | 35.77 | 42 | 34.15 |
| | General Commercial Area | 6,670,000 | 84 | 12.59 | 6 | 0.9 | 236 | 35.38 | 341 | 51.12 |
| | Neighborhood Commercial Area | 240,000 | 6 | 25 | 1 | 4.17 | 13 | 54.17 | 4 | 16.67 |
| | Distribution Commercial Area | 900,000 | 36 | 40 | 4 | 4.44 | 33 | 36.67 | 17 | 18.89 |
| | Industrial Area | 34,300,000 | 2269 | 66.15 | 138 | 4.02 | 773 | 22.54 | 250 | 7.29 |
| | Green Area | 356,260,000 | 33,270 | 93.39 | 443 | 1.24 | 1489 | 4.18 | 424 | 1.19 |

In order to apply measures in terms of urban planning to mitigate urban flood damages, it is necessary to establish an objective and reasonable method of setting dangerous areas, such as disaster prevention areas, and this study emphasizes that measures for disaster prevention shall be established by dividing the results of the urban flood risk analysis into 100 m × 100 m geographic grids and primarily considering the areas with a relatively high ratio of red zones.

## 4. Conclusions

The development and expansion of the city have brought about a variety of environmental problems, and the disaster in the city is gradually becoming larger and diversified due to the influence of climate change. Under this circumstance, the importance of urban disaster prevention has been emphasized, and this study focused on flooding, which accounts for a significant portion of disasters related to climate change. This study intended to contribute to the reduction and minimization of flood damage in the case of heavy rainfall in urban areas, which are concentrated with population and major facilities, by increasing the urban spatial efficiency through the grading of flood risk. This study developed and applied a flood risk assessment model to Changwon City and obtained the results described below.

First, as an assessment model, an objective and scientific evaluation model which reflects urban spatial concepts and characteristics of buildings was developed by applying an urban flood risk assessment model. Additionally, the non-structural characteristics were identified and the land use and vulnerability of building units were analyzed to minimize damage from natural disasters as a long-term measure. This study intended to construct an urban spatial model to primarily apply to less vulnerable and risky areas and suggested a new paradigm for the integrated study of 'environment + city' rather than individual planning or the study of environmental planning and urban planning. The developed model provides urban planners with a flood risk map when developing flood risk management strategies to ensure that a strong decision is made on flood management options such as land use, which represents a major challenge for flood risk, and disaster management challenges faced by climate change. Therefore, the ability of urban planners to make such decisions is important in reducing the social, economic, and physical (infrastructure) impact of flood damage.

Second, a set of indicators with non-structural characteristics that are related to land use and building characteristics were derived to assess vulnerability. The priority for each 100 m × 100 m geographic grid for flood damage was determined based on the fuzzy analytical values for each indicator: officially assessed land price, floor area ratio, underground area, decline of building, and material of building. When the same area was flooded, the area with the largest flood damage in a social and economic sense, which is the area with high vulnerability, was the area with the highest development density, including the central commercial area. The areas that had key functions in the city in the past, as the old town center, although currently declining, demonstrated high vulnerability.

Third, an urban flooding map was constructed by using the HEC-Ras model based on the basic environmental factors, such as rainfall data and topographic data, which were developed by the

Ministry of the Interior and Safety. The results of urban flooding area evaluation confirmed that the flood depth of the northern area adjacent to the river and reservoir is high. The flood depth was also high in the areas adjacent to Changwon City Hall, which play key roles in Changwon City, the target area, and which are also considered as the town center.

Fourth, the flood risk was analyzed by overlapping the results of vulnerability analysis and exposure analysis. As a result of analyzing the risk of urban flooding by four grades: green, yellow, orange, and red zones, it was confirmed that red zones were formed centering on the central commercial areas, where a variety of infrastructures, including public institutions with key functions in the city, were developed. This is related to the development of the economic, convenience, and efficient aspects of the city. The orange zone has a fan-shaped distribution centering on the red zone, which also shows the tendency of branching out from major functions and socio-economic parts of the city, and by using districts, they were classified into distribution commercial areas, including general commercial areas and semi-residential areas. The distribution of the yellow zone centered on the residential area, confirming that it is distributed in the area that was formed a long time ago. It is deemed that this is closely related to the aging of the buildings. The green zone was mainly distributed in green areas, where the environmental values or the environment rather than the value of development land use need to be preserved.

Lastly, the analysis of the urban flood risk in each administrative district showed that Euichang-gu (13.07%) had the highest urban flood risk with the highest percentage of red zone in Changwon, followed by Seongsan-gu (4.07%), Masan Hoewon-gu (2.68%), Masan Happo-gu (1.87%), and Jinhae-gu (1.78%), in respective order. This was attributed to the generally high property values and building density in the commercial areas of Seongsan-gu and Euichang-gu, which represent the central region for new town development. In other words, the land use plan should be established by allocating the green zone with the lowest urban flood risk and other use districts appropriately and disaster prevention facilities and space facilities that can reduce flooding should be allocated appropriately to prepare measures primarily for the areas with high risk. However, the difference in the analysis results of the five districts, which differed from each other, indicates that the degree of damage from disasters may vary according to the local environment and characteristics, newly developed town areas, the decline of the old town center and geographical location.

**Author Contributions:** K.P. (Kiyong Park) conceptualized the research, performed the formal analysis, and wrote the first draft of the paper. M.H.L. (Man-Hyung Lee) provided feedback on the research approach, and reviewed the first draft of the paper. All authors revised the paper and agreed on the final version of the paper.

**Acknowledgments:** This paper was financially supported by Ministry of the Interior and Safety as "Human resource development Project in Disaster management".

**Conflicts of Interest:** The authors declare no conflict of interest.

## References

1. Chang, L.C.; Chang, F.J.; Yang, S.N.; Kao, I.F.; Ku, Y.Y.; Kuo, C.L.; bin Mat Amin, I.M.Z. Building an Intelligent Hydroinformatics Integration Platform for Regional Flood Inundation Warning Systems. *Water* **2019**, *11*, 9. [CrossRef]
2. Lytle, D.A.; Poff, N.L. Adaptation to natural flow regimes. *Trends Ecol. Evol.* **2004**, *19*, 94–100. [CrossRef] [PubMed]
3. Tockner, K.; Lorang, M.; Stanford, J.A. River flood plains are model ecosystems to test general hydrogeomorphic and ecological concepts. *River Res. Appl.* **2010**, *86*, 76–86. [CrossRef]
4. Miguez, M.G.; Veról, A.P.; de Sousa, M.M.; Rezende, O.M. Urban Floods in Lowlands—Levee Systems, Unplanned Urban Growth and River Restoration Alternative: A Case Study in Brazil. *Sustainability* **2015**, *7*, 11068–11097. [CrossRef]
5. Tingsanchali, T. Urban flood disaster management. *Procedia Eng.* **2012**, *32*, 25–37. [CrossRef]
6. Walsh, R.; Davies, H.; Musa, S. Flood frequency and impacts at Khartoum since the early nineteenth century. *Geogr. J.* **1994**, *160*, 266–279. [CrossRef]

7. Zevenbergen, C.; Veerbeek, W.; Gersonius, B.; van Herk, S. Challenges in urban flood management: Travelling across spatial and temporal scales. *J. Flood Risk Manag.* **2008**, *1*, 81–88. [CrossRef]
8. Liao, K.; Le, T.A.; van Nguyen, K. Urban design principles for flood resilience: Learning from the ecological wisdom of living with floods in the Vietnamese Mekong Delta. *Landsc. Urban Plan.* **2016**, *155*, 69–78. [CrossRef]
9. Dihn, Q.; Balica, S.; Popescu, I.; Jonoski, A. Climate change impact on flood hazard, vulnerability and risk of the Long Xuyen Quadrangle in the Mekong Delta. *Int. J. River Basin Manag.* **2012**, *10*, 103–120.
10. Selenica, A.; Kuriqi, A.; Ardicioglu, M. Risk assessment from floodings in the rivers of albania. In Proceedings of the International Balkans Conference on Challenges of Civil Engineering, Tirana, Albania, 23–25 May 2013.
11. Kang, L.; Jiang, S.; Hu, X.; Li, C. Evaluation of Return Period and Risk in Bivariate Non-Stationary Flood Frequency Analysis. *Water* **2019**, *11*, 79. [CrossRef]
12. Adhikari, P.; Hong, Y.; Douglas, K.R.; Kirschbaum, D.B.; Gourley, J.; Adler, R.; Robert Brakenridge, G. A digitized global flood inventory (1998–2008): Compilation and preliminary results. *Nat. Hazards* **2010**, *55*, 405–422. [CrossRef]
13. Güneralp, B.; Güneralp, I.; Liu, Y. Changing global patterns of urban exposure to flood and drought hazards. *Glob. Environ. Chang.* **2015**, *31*, 217–225. [CrossRef]
14. Eissa, A.E.; Zaki, M.M. The impact of global climatic changes on the aquatic environment. *Procedia Environ. Sci.* **2011**, *4*, 251–259. [CrossRef]
15. Broekx, S.; Smets, S.; Liekens, I.; Bulckaen, D.; de Nocker, L. Designing a long-term flood risk management plan for the Scheldt estuary using a risk-based approach. *Nat. Hazards* **2011**, *57*, 245–266. [CrossRef]
16. Yin, J.; Ye, M.; Yin, Z.; Xu, S. A review of advances in urban flood risk analysis over China. *Stoch. Environ. Res. Risk Assess.* **2015**, *29*, 1063–1070. [CrossRef]
17. Zope, P.E.; Eldho, T.I.; Jothiprakash, V. Impacts of urbanization on flooding of a coastal urban catchment: A case study of Mumbai City, India. *Nat. Hazards* **2014**, *75*, 887–908. [CrossRef]
18. Garcia, E.; Loáiciga, H. Sea-level rise and flooding in coastal riverine flood plains. *Hydrol. Sci. J.* **2014**, 37–41. [CrossRef]
19. Kim, H.; Lee, D.; Sung, S. Effect of Urban Green Spaces and Flooded Area Type on Flooding Probability. *Sustainability* **2016**, *8*, 134. [CrossRef]
20. Liu, W.; Chen, W.; Peng, C. Assessing the effectiveness of green infrastructures on urban flooding reduction: A community scale study. *Ecol. Model.* **2014**, *291*, 6–14. [CrossRef]
21. Gaur, A.; Gaur, A.; Yamazaki, D.; Simonovic, S.P. Flooding Related Consequences of Climate Change on Canadian Cities and Flow Regulation Infrastructure. *Water* **2019**, *11*, 63. [CrossRef]
22. Intergovernmental Panel on Climate Change (IPCC). *Climate Change 2007: Impact, Adaptation and Vulnerability, Fourth Assessment Report*; Cambridge University Press: Cambridge, UK, 2007.
23. Moss, R.H.; Brenkert, A.L.; Malone, E.L. *Vulnerability to Climate Change*; U.S. Department of Energy: Oak Ridge, TN, USA, 2001.
24. The United Nations Educational, Scientific and Cultural Organization (UNESCO). *International Hydrological Programme (IHP), Water Security: Responses to Local, Regional, and Global Challenges, Strategic Plan VIII (2014–2021)*; United Nations Educational, Scientific and Cultural Oranization: Paris, France, 2013.
25. Mehr, R.I.; Hedges, B.A. *Risk Management: Concepts and Applications*; Richard Irwin Inc.: Homewood, CA, USA, 1974.
26. Shin, S.Y.; Kim, H.R. *Analyzing Relationships between Land Use Characteristics and Flood Damage Areas*; Seoul Development Institute: Seoul, Korea, 2011.
27. Burton, I.; Kates, R.W.; White, G.F. *The Environment as Hazard*, 2nd ed.; The Guilford Press: New York, NY, USA, 1993.
28. Norman, B. Principles for an Intergovernmental Agreement for Coastal Planning and Climate Change in Australia. *Habitat Int.* **2008**, *33*, 293–299. [CrossRef]
29. United Nations Environment Programme (UNEP). *Mainstreaming Climate Change Adaptation into Development Planning: A Guide for Practitioners*; UNEP: Nairobi, Kenya, 2012.
30. Hideki, K. Urban Disaster Prevention. Tokyo, Japan, 2011. Available online: https://www.urd.org/IMG/pdf/Urban_Preparedness_Kathmandu_Full_Report_.pdf (accessed on 7 May 2019).
31. Park, K.Y.; Lee, M.H. Vulnerability analysis of urban district on the urban flood damage: A case study Changwon. *Desalin. Water Treat.* **2018**, *119*, 27–35. [CrossRef]

32. Lee, C.H. *Development of the Regional Safety Assesment Model in Seoul-Focusing on Flood*; Seoul Development Institute: Seoul, Korea, 2006; pp. 1–130.
33. Han, H. *Climate Change Impact Assessment and Development of Adaptation Strategies in Korea*; Korea Environment Institute: Sejong, Korea, 2007; pp. 1–417.
34. Park, H. *Disaster Site Investigation and Disaster Prevention Planning Standard Establishment for Urban Space and Facilities*; National Disaster Management Research Institute: Ulsan, Korea, 2010; pp. 1–371.
35. Jang, M.S. Vulnerability Analysis of Building Inundation caused by Storm Surge in Masan Bay. Master's Dissertation, Sungkyunkwan University, Seoul, South Korea, 2015; pp. 1–55.
36. Unanwa, C.O.; McDonald, J.R.; Mehta, K.C.; Smith, D.A. The Development of Wind Damage Bands for Buildings. *J. Wind Eng. Ind. Aerodyn.* **2000**, *84*, 119–149. [CrossRef]
37. Zadeh, L.A. Fuzzy sets. *Inf. Control* **1965**, *8*, 338–353. [CrossRef]
38. Chang, N.; Chen, H.W.; Ning, S.K. Identification of river water quality using the Fuzzy Synthetic Evaluation approach. *J. Environ. Manag.* **2001**, *63*, 293–305. [CrossRef] [PubMed]
39. Gopal, S.; Woodcock, C. Remote sensing of forest change using artificial neural networks. *IEEE Trans. Geosci. Remote Sens.* **1996**, *34*, 398–404. [CrossRef]
40. Curlander, J.; Kober, W. Rule Based System for Thematic Classification in SAR Imagery. In Proceedings of the International Geoscience and Remote Sensing Symposium (IGARSS' 92), Houston, TX, USA, 26–29 May 1992; pp. 854–856.
41. Tsatsoulis, C. Expert systems in remote sensing applications. *IEEE Geosci. Remote Sens. Newsl. June* **1993**, 7–15.
42. Benz, U.C.; Hofmann, P.; Willhauck, G.; Lingenfelder, I.; Heynen, M. Multi-resolution, object-oriented fuzzy analysis of remote sensing data for GIS-ready information. *J. Photogramm. Remote Sens.* **2004**, *58*, 239–258. [CrossRef]
43. Ministry of Land, Infrastructure and Transport. *A Study on the Effective Reducing for Prevention the Urban Flood Damage*; Ministry of Land, Infrastructure and Transport: Sejong, Korea, 2008.
44. Changwon-City. 2025 Urban Masterplan for Changwon. Available online: www.changwon.go.kr (accessed on 30 April 2019).
45. Ministry of the Interior and Safety. *Insurance Rate Making and Mapping Based on Natural Disaster Risk*; Ministry of the Interior and Safety: Sejong, Korea, 2017.
46. Kron, W. *Flood Risk = Hazard × Exposure × Vulnerability, Flood Defence*; Science Press: New York, NY, USA, 2002; pp. 82–97.
47. Brooks, N.; Adger, W.N.; Kelly, P.M. The determinants of vulnerability and adaptive capacity at the national level and the implications for adaptation. *Glob. Environ. Chang.* **2005**, *15*, 151–163. [CrossRef]

water

MDPI

Article

# Built-Up Growth Impacts on Digital Elevation Model and Flood Risk Susceptibility Prediction in Muaeng District, Nakhon Ratchasima (Thailand)

**Patiwat Littidej * and Nutchanat Buasri**

Department of Geo-Informatics, Faculty of Informatics, Mahasarakham University, Mahasarakham 44150, Thailand

* Correspondence: patiwat.l@msu.ac.th

Received: 11 June 2019; Accepted: 12 July 2019; Published: 18 July 2019

**Abstract:** The transformation of land-use and land cover in Nakhon Ratchasima province, Thailand has rapidly changed over the last few years. The major factors affecting the growth in the province arise from the huge expansion of developing areas, according to the government's development plans that aim to promote the province as a central business-hub in the region. This development expansion has eventually intruded upon and interfered with sub-basin areas, which has led to environmental problems in the region. The scope of this study comprises three objectives, i.e., (i) to optimize the Cellular Automata (CA) model for predicting the expansion of built-up sites by 2022; (ii) to model a linear regression method for deriving the transition of the digital elevation model (DEM); and (iii) to apply Geographic Weighted Regression (GWR) for analyzing the risk of the stativity of flood areas in the province. The results of this study show that the optimized CA demonstrates accurate prediction of the expansion of built-up areas in 2022 using Land use (LU) data of 2-year intervals. In addition, the predicting model is generalized and converged at the iteration no. 4. The prediction outcomes, including spatial locations and ground-water touch points of the construction, are used to estimate and model the DEM to extract independent hydrology variables that are used in the determination of Flood Risk Susceptibility (FRS). In GWR in the research called *FRS-GWR*, this integration of quantitative GIS and the spatial model is anticipated to produce promising results in predicting the growth and expansion of built-up areas and land-use change that lead to an effective analysis of the impacts on spatial change in water sub-basin areas. This research may be beneficial in the process of urban planning with respect to the study of environmental impacts. In addition, it can indicate and impose important directions for development plans in cities to avoid and minimize flood area problems.

**Keywords:** flood risk susceptibility; *FRS-GWR* modeling; built-up growth prediction; Thailand

## 1. Introduction

Land-use in areas typically involves dynamical processes that develop land which changes over time, based on the evolution of the economy, society, and population in the areas [1]. Population growth results and increases the demands and expansion of land-use in such areas. As a consequence, this fabricates land-use planning (especially the expansion of residential and building areas), which leads to substantial replacements owning to restricted resources [2–4]. For example, there is an inevitable rational in some areas to replace dwelling and development zones in agricultural lands [5]. In the last five years, the increased development and expansion of built-up areas has impacted dramatically on land-use and land-cover characteristics. In Nakhon Ratchasima (urban zone), Thailand, large areas of agricultural land have essentially been transformed for development, housing, and dwelling zones according to the current government's master plans. The plans aim to increase the potential transportation capabilities in order to cope with the expansion of the cities in the country by launching a number of construction projects in the province, i.e., a Highway (no. 6), a double-track train line, and a high-speed train project [6]. As a result, this eruptive non-planned land-use change can cause significant environmental problems, for instance, water logging, traffic, and buildup of solid waste [7–9].

Land-use change is one of the factors that can impact on flow accumulation in water networks, including flow accumulation, flow length, and areas of sub-basins. These flow accumulations are sensitive hydrological factors that play a vital role in determining the change of the DEM in the areas [10,11]. The impact of urban growth is one of the significant land-use changes affecting surface runoff within the catchment area [12]. Urban expansion also leads to the removal of trees and vegetation, causing a decrease in evapotranspiration. The construction of roads and culverts has effects which may include reduction of infiltration, decline in the groundwater table, increased surface runoff, and a reduction of base flows [13–15]. Built-up areas lead to an increase in DEM surfaces which in turn reduces the runoff concentration time. Accordingly, higher peak discharge rates occur sooner after rainfall in the catchment area [16]. In addition, the runoff volume and potential flood risk greatly increases [17–19]. Converting agricultural land to built-up arcades, for example, may require land-fill work to level the height of the land up to at least 2–3 m before construction can be carried out to minimize water logging risks. This land-fill can ultimately interfere and block the flow accumulation, thus incurring a change in the flow directions. As a result, rain-water cannot be drained to main local rivers effectively when a large volume of rain is received; this subsequently creates water logging in the area [20–22]. Land height in sub-basin regions is an important factor that imposes a substantial flow length and flow accumulation. If the density of flow length and the number of accumulated flows are small compared with the size of the sub-basins, the performance of the water drain process will become degraded and result in water-logging [23,24]. Prediction of the growth and expansion of these development areas can be utilized to generate an accurate transition DEM map [25,26]. In addition, it can be used to analyze quantitative hydrological factors. However, land-use modeling is more complicated as it needs to integrate different pieces of contextual information, such as spatial information and some independent variables that drive land-use change [27–31]. The factors that make DEM different in each area are also dependent on the nature of the style of the land-use, and the social and economic form. In recent studies, there has been no research which has been used to increase the height of the area with such local factors.

Predicting land-use and land-cover change using mathematic models and geo-information systems is one of the popular methods as it can visualize outcomes in spatial information patterns, which are not limited to the quantitative information available in other conventional methods. Land-use information is fundamental to analyzing the changes of land coverage by analyzing the various spatial software [31]. This work applies a wide range of spatial models, such as Geomod2 [32], CA, SLEUTH Urban Growth [33], Lucas [34], and LTM [35]. Each model has its own characteristics and processes. However, they function in the same capacities, which are used to model the change of land- use and land-cover. Additionally, each technique is applicable to different geo-information software. CLUE-S can be used together with ArcGIS, QGIS, and IDRISI. Lucas, on the other hand, can be coupled with GRASS and Geomod2, which can be applied with IDRISI. Likewise, CA is a sub- module available in IDRISI. Therefore, there are a number of tools that can be applied to model the change of land-use and land-cover. Integration between land-use forecast models and the DEM modeling can be applied to the modeling to analyze the risk of flood disaster.

This research applies CA integrated with optimization schemes to predict built-up growth in areas that have experienced rapid growth (construction and development). The prediction model is used, thereafter, to model the change of DEM in order to precisely locate and estimate the change of DEM according to the growth of buildings in the areas. In addition, it is utilized in modeling GWR for analyzing sensitive areas and areas that are at risk to water flooding resulting from the growth of construction in these areas. This integration of quantitative and spatial models is anticipated to produce promising results in predicting spatial change in water sub-basin areas. This study is aimed at

(i) optimizing the CA model for predicting the expansion of built-up polygon in 2022; (ii) modeling a linear regression method for deriving the transition of DEM and hydrological characteristic analysis for independent variable extraction; and (iii) applying the GWR model for analyzing the risk of the stativity of flood areas.

## 2. Research Methodology

### *2.1. Study Area*

Amphoe Mueang Nakhon Ratchasima, Thailand has an area of approximately 645.84 sq. km There are 22 communities in the city, "Korat", which is located in the center of the province. In 2014, land-use and land-cover in the city was deliberately divided into five main types (as illustrated in Figure 1 and Appendix A Table A1), i.e., (i) agricultural areas (385.01 sq. km 59.61%); (ii) forest (7.12 sq. km, 1.10%); (iii) water resources (7.45 sq. km, 1.15%); (iv) built-up and dwellings (202.68 sq. km, 31.40%); and (v) others (43.58 sq. km, 3.33%) respectively. Geographically, this study area is located in the Lamtaklong basin (WGS 1984 UTM Zone 48 N (X = 169,957 to 202,237) E and (Y = 1,637,208 to 1,675,772) N). Spatial level is between 164 to 277 m. Lamtaklong River is the Main River and cuts through the center of the province. This river is the major source that supports the water- draining flowing of the floodway from west to east across the province. From a hydrological perspective, the majority of areas in the northeast are lowland areas, which sit lower than those in the east. As the city is situated close to Lamtaklong River and there is an ongoing construction project in the area, including the endpoint of the highway and bypass road (Nakhon Ratchasima–Khon Kean) congestion, the area is susceptible to a water-logging problem. In some situations, water cannot drain as run-off efficiently after rainfall for 20 min. This non-planned expansion of built-up and developing areas causes land-filling (of up to 2 m from regular levels), and a change of land-use conditions, resulting in tremendous problems with the sub-basins and from flow accumulation.

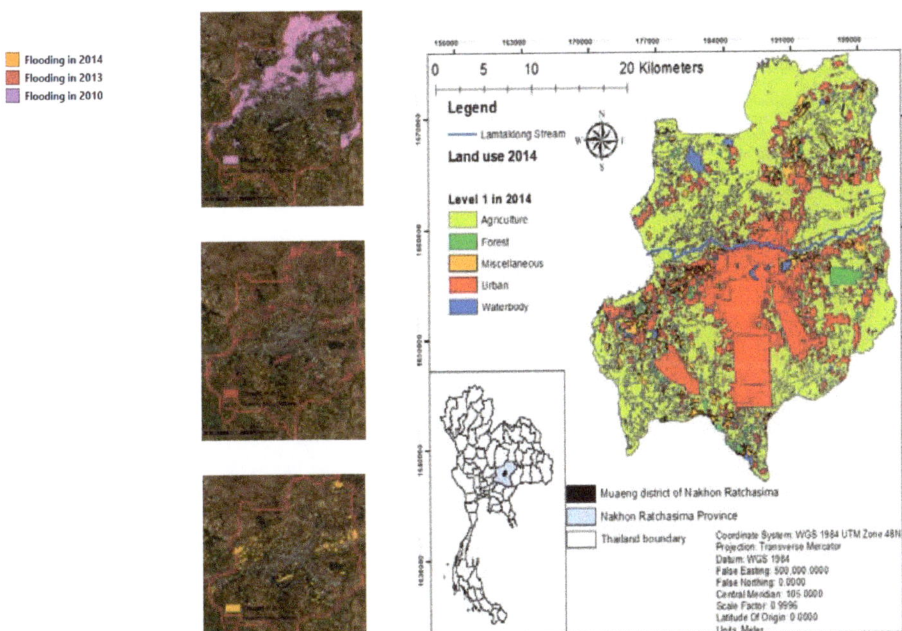


**Figure 1.** Significant floods in 2010, 2013, and 2014 covered in land-use/land cover (LU/LC) of Mueang district extent, Nakhon Ratchasima province derived using visual interpretation in December 2014.

To alleviate these problems, therefore, this research devised three objectives (depicted in Figure 2) as follows. Firstly, the prediction of built-up areas by 2022 using CA is performed using high-resolution IKONOS data (downloaded from Google Earth Pro). The image data was collected in December 2014, 2016, and 2018, as these times are at the end of the annual land-use change with less cloud scattering hindering the visualization of the data. The model will project predicted data to produce a prediction map of built-up and development areas by 2022. However, this predictive model cannot be appropriately applied to this study area due to the fact that the change of land-use and the expansion of built-up and construction in the area is exponentially eruptive-resulting from a number of the government's mega projects. Therefore, optimization techniques are implemented over CA.

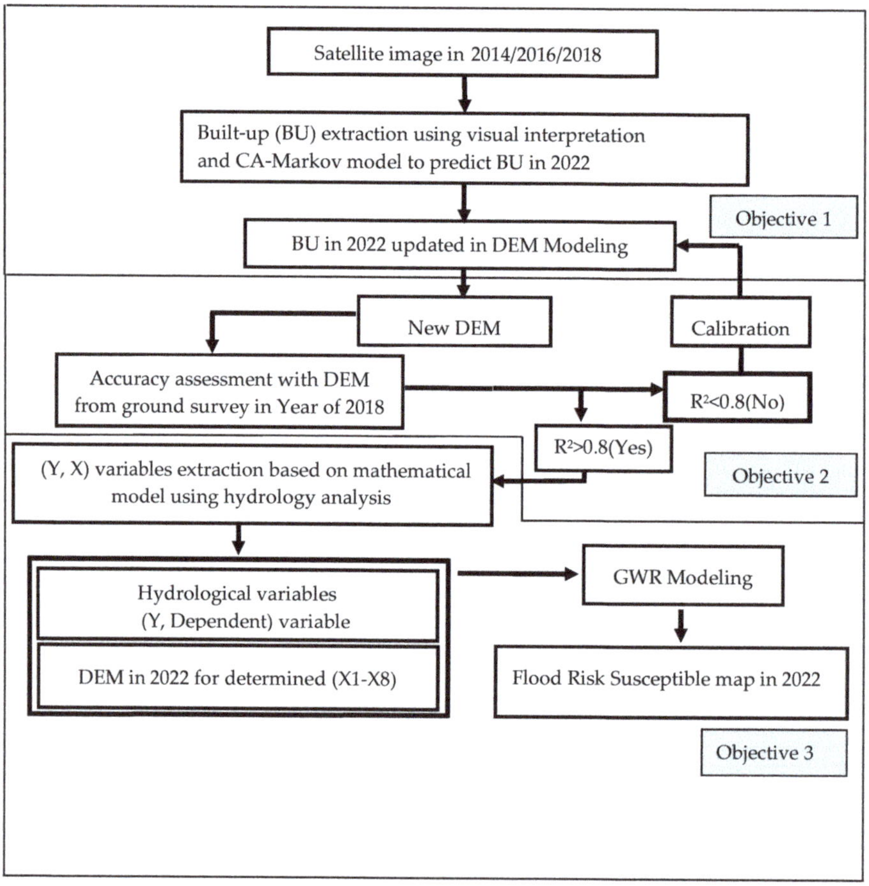


**Figure 2.** Framework of the study.

## 2.2. *Research Approach*

The procedure for the main operation has three phases: 1. Translation of land data from 3 satellite image data over time, then taking the data to forecast land-use in 2022. The next step is to bring out the projected year land utilization data from the model to calibrate and use to generate numerical height information. This process uses linear models and independent variables that involve the height in the study area. The final step is the introduction of the projected altitude data based on the change from the building's rise as a basis to create a combination of variables based on the independent variables in the GWR model. This thorough research process can be described in spatial data as follows:

An iterative procedure was applied to generate a number of built-up features in the study area. Their polygon was used in the new DEM modeling. Secondly, a new DEM manipulation was generated from the built-up area in 2022 via DEM linear modeling. The DEM prediction model was adjusted to have an accuracy greater than 0.8, and after that it is used as DEM in 2022. DEM in the base year and Euclidean distance of built-up to streamline are used in the model. Third, the updated heights are used to analyze the hydrologic characteristics to create a flood areas index (FAI) and Flood Risk Susceptible (FRS) model. The DEM in 2022 can generate independent variable data in eight different types of sub-basin. The GWR model tests the ability to predict FRS from the indexes $R^2$ and AIC. The factors are screened only using the independent variables that influence the sensitive area and the risk of flooding from the increased number of buildings.

### 2.3. Built-Up Growth Prediction Model

Many studies and researches on prediction of urban growth using different models have been carried out throughout the world. In this study, we select two models for urban growth prediction: The Cellular Automata (CA) model using CA-Markov in IDRISI software package and the Geographic weighted regression (GWR) model which are summarized herein. CA and GWR models are selected to predict built-up growth, the results of which are then compared with the interpreted built-up polygon in 2018.

In references by Benenson and Suwit [36,37] it was said, after that, the model which provides greater accuracy will be used for built-up growth prediction in 2022. The main tasks concern urban growth prediction in 2022 using CA-Markov, selection of the optimum iteration model for built-up growth prediction in the future and prediction of urban growth in 2018 and 2022 as well as similar approaches conducted by Benenson and Suwit [36,37]. Cellular automata are dynamic models discrete in time, space and state. A simple cellular automata A is defined by a lattice ($L$), a state space ($Q$), a neighborhood template $\delta$ and a local transition function ($f$):

$$A = (L,\ Q,\ \delta,\ f) \tag{1}$$

Each cell of $L$ can be in a discrete state out of $Q$ according to Benenson and Suwit [36,37]. The cells can be linked in different ways. Cells can change their states in discrete time-steps. Usually, cellular automata are synchronous, i.e., all cells change their states simultaneously. The fate of a cell is dependent on its neighborhood and the corresponding transition function $f$ [38,39]. The formal definition of Markov processes is very close to that of CA. The Markov process is considered in discrete time and characterized by variables that can be in one of $N$ states from $S = \{S1, S2, \ldots, S_N\}$. The set $T$ of transition rules is substituted using a matrix of transition probabilities ($P$) and this is reflective of the stochastic nature of the process:

$$P_{ij} = \begin{bmatrix} P_{11} & P_{12} & P_{1n} \\ P_{21} & P_{22} & P_{2n} \\ \cdots\cdots\cdots & \cdots\cdots\cdots & \cdots\cdots\cdots \\ P_{n1} & P_{n2} & P_{nn} \end{bmatrix} \tag{2}$$

where $P_{ij}$ is the conditional probability that the state of a cell at moment $t + 1$ will be $S_j$, given it is $S_i$ at moment $t$:

$$prob(S_i \rightarrow S_j) = P_{ij} \tag{3}$$

Ref. by Suwit [37] said that the Markov process as a whole is given by a set of status $S$ and a transition matrix $P$. By definition, in order to always be 'in one of the states' for each $i$, the condition $\sum_j P_{ij} = 1$ should hold [39]. The research made the appropriate configuration of the iteration of CA model in IDRISI for built-up prediction by 2022.

### 2.4. Geographically Weighted Regression (GWR) Model of Flood Risk Susceptible (FRS) Prediction

Geographically Weighted Regression [40,41] is used to combine data in each point of observation into a regression model using a series of distance-related weights. The relationship between growth built-up area and location characteristics for a particular point, for example, is given a higher weight than for points further away from that point. The GWR regression model for Flood area predicted ($F_j$) to set as dependent variable, ($Y$) prediction in Nakhon Ratchasima municipality is shown in the following Equation (4).

$$F_j(u_iv_i) = \beta_0(u_iv_i) + \beta_1X_1(u_iv_i) + \beta_2X_2(u_iv_i) + \beta_3X_3(u_iv_i) + \\ \beta_4X_4(u_iv_i) + \beta_5X_5(u_iv_i) + \beta_6X_6(u_iv_i) + \beta_7X_7(u_iv_i) + \beta_8X_8(u_iv_i) + \varepsilon_i \quad (4)$$

where $\beta_0$ is the intercept term, $\beta_1, \beta_2, \beta_3, \beta_4, \beta_5$, and $\beta_6$ are spatially varying coefficients of the Normalized Digital Elevation Model Index (*NDEMI*, $X_1$), Normalized Contour Index (*CLI*, $X_2$), Area of Built-up Index (*ABUI*, $X_3$), Density of Built-up ($D_j$, $X_4$), Curvature Index (*CI*, $X_5$), Slope length Index (*SLI*, $X_6$), perimeters ($P_j$, $X_7$), and sub-basin area ($S_j$, $X_8$) attributes respectively, and $\varepsilon_i$ is an error term at point $i$, ($u_iv_i$) representing the coordinates of the $i^{th}$ point in study extent [41]. The GWR model is modeled using the GWR 3.0 software package which allows the use of a variety of calibration techniques to specify regression weights and to optimize bandwidth parameters. In this study, a fixed defined kernel with a bi-square function (in which the bandwidth was determined through the minimization of the Akaike Information Criterion (AIC) [40,41] was used. The reason for this is that the points in the spatial unit of analysis used are in regular and equal sizes. Monte Carlo tests [40,41] were also carried out to set the significance of the spatial variability in the local parameter estimates. Independent variable weight values are indicators of which independent variables will influence flood risk. The coefficients of independent variables have both positive and negative relationships affecting the change of the flood areas. A series of pre-tested, independent variables provides a satisfactory result in terms of statistical index value, which is defined as an independent variable to use the Flood Risk Susceptibility (FRS) prediction in each sub-basin.

The map showing the flood risk area size may not reflect the severity of the spatial unit of each sub-basin. Therefore, it is necessary to create an additional model to describe the sensitivity of the flood. Flood Area Index (FAI) was used as a major input in the Flood Risk Susceptibility (FRS) model and shown as a probability in the term between 0–100 in the following Equation (5) as FAI and Equation (6) as FRS, respectively.

Many of the previously flooded areas act as indicators of the frequency and severity of flooding in those areas. Flood information in the past was interpreted from the satellite imagery by Geo-Informatics and Space Technology Development Agency (Public Organization), (GISTDA). Flood data from 2010–2016 was able to be used to establish a Flood Area Index (FAI Dependent variable, (Y)) as shown in Equation (5)

$$F_j = \frac{\sum_{i=1}^{n} F_{ijk}}{N_k} \quad (5)$$

where $F_j$ is Flood Area Index of each sub-basin any $j$, $\sum_{i=1}^{n} F_{ijk}$ is the summation of flooding in the cell any $j$ of each sub-basin any $j$ in the year any $k$ and $N_k$ is a number of flooding the cell in the year any $k$.

$$FRS_j = \frac{F_j}{S_j} \times 100 \quad (6)$$

where $FRS_j$ is flood risk susceptibility in 2022 of each sub-basin any $j$ where the closer the value is to 1, the greater the risk (ranged as 0–1) and $F_j$ is Flood Area Index (FAI) predicted in each sub-basin of any $j$ (sq. km) and $S_j$ is the area of each sub-basin of any $j$ (sq. km).

### 2.5. Independent Variables Modeling from Hydrological Characteristics

This study begins by undertaking a Digital Elevation Model (DEM) modeling of a numerical height to predict the height of the space in the desired future year. These height data are extracted as an index of the independent variables to be used to create the GWR model. This also uses the flood data in the past that occurred in the leading area index of Flood Area Index (FAI). The FAI index is a dependent variable in the GWR model and is extracted to a boundary of each sub-basin and displays the details of the various indices.

#### 2.5.1. DEM Prediction Modeling

A new DEM in each sub-basin of each following year is derived from the linear model following Equation (6). The height of DEM on the base-year and the distance of built-up polygons from the stream are input into the model. The altitude change in the study area is a major factor affecting the independent variables mentioned above. The coefficient of the independent variable is used to identify the relationship between the distances from the stream to the predicted altitude as shown in Equation (7)

$$D_{i(n+1)} = D_{i(n)} \pm L_{is} \tag{7}$$

where $D_{i(n+1)}$ is elevated point at any $i$ in next ($n$ + 1) year (meters), $D_{i(n)}$ is elevation point of cell any $i$ in $n$ year (meters,) and $L_{is}$ is a distance between the center point of built-up polygons of feature any $i$ to the closet Lamtaklong stream line feature $s$ (meters).

#### 2.5.2. Normalized Digital Elevation Model Index (NDEMI, X1)

The DEM index is made to analyze the difference in the average altitude in the sub-basin area, with the assumption that if any of the sub-basin areas are different, then there is a greater risk of flooding. However, the index only applies to this analysis of risk-prone areas. Flooding may not be able to identify the total area risk due to the difference in the altitude of the area, resulting in the water flowing into other subgroups and quickly venting. Therefore, the risk analysis of flooding is necessary to be used in conjunction with other independent variables. The output of this index is adjusted to a standard in the range of 0–1. The higher the altitude, the closer the value is to 1, and the lower value will approach 0. The index guideline makes it possible to measure the difference in the same standard of that sub-basin. The index shows the calculation of Equation (8).

$$NDEMI_j = \frac{\sum\limits_{i=1}^{n}\left[\frac{E_i - E_{min}}{E_{max} - E_{min}}\right]}{\sum\limits_{i=1}^{n} N_{ij}} \tag{8}$$

where the $NDEMI_j$ index is the normalized digital elevation model of the sub-basin any $j$, is the proportion of the standard average of the difference in height. $E_i$ is elevation point of any cell $i$, $E_{max}$ is the highest elevation point of each sub-basin any $j$ (meters), $E_{min}$ is the lowest elevation point of each sub-basin any $j$ (meters) and $\sum\limits_{i=1}^{n} N_{ij}$ is the summation of the number of elevation points of any cell $i$ in each sub-basin any $j$.

2.5.3. Normalized Contour Index (CLI, X2)

The Altitude line length index is calculated based on the multiples of the line length at different heights per size of each area in any sub-basin. A high index value represents the difference in the altitude of a basin. A lower index value indicates that the basin is smooth, as shown in Equation (9).

$$CLI_j = \frac{\sum_{i=1}^{n} C_k L_i}{P_j} \tag{9}$$

where $CLI_j$ is the normalized contour index of sub-basin any $j$. It is the proportion of the sum product of a height level and the length of contour lines per any perimeter of sub-basin any $j$, $C_k$ is a level of contour line any $k$ in sub-basin any $j$, $L_i$ is a length of contour line any $i$ of contour line any $k$ in sub-basin any $j$ and $P_j$ is the perimeter of sub-basin any $j$.

2.5.4. Area of Built-Up Index (ABUI, X3)

This is the proportion of the summation of the built-up area within sub-basin any $j$ per the extent of sub-basin any $j$ (sq. km). The index is used to analyze the larger and smaller built-up space. The increase in the building area makes the flow accumulation and the water flow direction in the basin area more likely to change than the area that rarely changes, as shown in Equation (10).

$$ABUI_j = \frac{\sum_{i=1}^{n} ABUI_i}{S_j} \tag{10}$$

where $ABUI_j$ is the area of built-up Index of sub-basin any $j$, $\sum_{i=1}^{n} ABUI_i$ is the summation of the built-up area in the polygon any $i$ of each sub-basin any $j$ and $S_j$ is the area of sub-basin any $j$ (sq. km).

2.5.5. Density of Built-Up ($D_j$, X4)

This index is similar to the indexes that are referenced above. However, the analysis of the water path change from the built-up area alone may not be sufficient to predict the flood risk susceptible area. In addition, the number of buildings is rapidly increasing and the boundaries are becoming close to one another, causing the water flow to be discontinued and the water is subsequently becoming trapped in a sub-watershed area that looks like this. Using this independent variable with an index (*ABUI*), it is possible to refer to the level of growth of buildings associated with flooding in the sub-basin, as shown in Equation (11).

$$D_j = \frac{\sum_{i=1}^{n} N_i}{S_j} \tag{11}$$

where $D_j$ is built-up density of sub-basin any $j$, $N_i$ is a number of built-up areas at the center point of polygon any $i$ within sub-basin any $j$ and $S_j$ is the area of sub-basin any $j$ (sq. km).

2.5.6. Curvature Index (CI, X5)

In the calculation of areas of the surface curvature, the convex is curved, textured, inverted, or has an embossed appearance, and the concave has a curved, textured, rounded appearance or a puddle. Convex-style areas pose a risk of less flooding than a concave surface, if the area appears flat, there is a value of two areas approaching the center. The index value of any basin is even more vulnerable and susceptible to flooding less than the low index area, as shown in Equation (12).

$$CI_j = \frac{\sum_{i=1}^{n} A_{convex(i)}}{S_j} \tag{12}$$

where $CI_j$ Curvature Index of sub-basin any $j$, $A_{convex(i)}$ is the area of Convex-style areas of polygon any $i$ within sub-basin any $j$ (sq. km) and $S_j$ is the area of sub-basin any $j$ (sq. km).

2.5.7. Slope Length Index (SLI, X6)

The index is calculated from the proportion of the sum of the slope length per each perimeter in the sub-basin. This index is used to analyze high-index values, the high and long slopes in the sub-basin area, allowing water to accumulate in the lowest possible areas while the risk of detention and drainage is low, as shown in Equation (13).

$$SLI_j = \frac{\sum_{i=1}^{n} SL_i}{P_j} \tag{13}$$

where $SLI_j$ is Slope length Index of sub-basin any $j$, $SL_i$ is a length of slope of polyline any $i$ within sub-basin any $j$ (km) and $P_j$ is the perimeter of sub-basin any $j$.

In addition to the previously-displayed indexes, there is also an independent variable that is used as an important piece of data to determine the index values, such as the perimeters where ($P_j$, $X_7$) is a calculation of the length of the circumference of the sub-basin, and the sub-basin area where ($S_j$, $X_8$) is a calculation of the area of the sub-basin, where both of these values can be calculated from the mathematical principles of the spatial relationship.

## 3. Results and Discussion

### 3.1. Optimal Iterations of CA-Markov Model for Built-Up Growth Prediction in 2022

The results of interpreting satellite imagery from the years 2014, 2016, and 2018 using the interpretation method interpreted visually (Visual interpretation) based on the composition of the interpretation consisting of the shape (Shape), size (Pattern), the intensity of colors, and colors (Tone and color), the texture (Shadow) location, and the association (Site) and its relevance are classified into five categories, including; (1) community areas and buildings: areas with all types of buildings and residential trade zones as well as government offices and transportation routes; (2) Miscellaneous areas including open space; (3) other agricultural areas such as areas, fields, garden areas, etc.; (4) water source areas that are both natural and man-made water sources; and (5) forest areas, including natural forest areas and planted forests. The results of the 2016 satellite image interpretation found that: (1) the Built-up area comprised a total of 209.45 sq. km (32.43%); (2) the miscellaneous area was 40.25 sq. km (6.25%); (3) the water source area was 7.06 sq. km (1.09%); (4) the forest area was 7.03 sq. km (1.08%); and (5) the agricultural area was 382.05 sq. km (59.15%). The effect of interpreting satellite imagery from 2018 found that: (1) the Built-up area was 225.88 sq. km (34.97%); (2) the miscellaneous area was 36.47 sq. km (5.65%); (3) the water source area was 6.91 sq. km (1.07%); (4) the forest area was 6.52 sq. km (1.01%); and (5) the agricultural area was 370.06 sq. km (57.30%), respectively, and all interpretation result information is shown in the Appendix A Tables A1–A10.

The transition matrix for built-up prediction in 2022 is shown in Table A10 and the dispersion position shown in Figure 3. The result of the 2022-year satellite image translation found that: (1) the Built-up area was 242.85 sq. km (37.6%); (2) the miscellaneous area was 27.49 sq. km (4.26%); (3) the water source area was 6.97 sq. km (1.08%); (4) the forest area was 6.22 sq. km (0.96%); and (5) the agricultural area was 362.31 sq. km (56.1%). Building and road areas tend to increase every year. The main types of soil cover that are transformed into building areas are: Agricultural areas and water sources mostly included in some miscellaneous areas. The water source area is being continually reduced due to the fact that the land is reclaimed in the original water source to make way for constructions, such as housing projects, housing, and other types of homes, as the position of change in this way is formed along the road, avoiding the city and the area near the end of the Highway (main road number-6). Miscellaneous areas are relatively constant, but there is a slight decline shown in the year 2018. This soil coverage data layer is converted into raster data in the form of a file extension (.RST) to import into a raster format. The forecast growth of built-up areas is conducted with CA-Markov models using the IDRISI version 15.0 to create a matrix of changes and analyze the optimal iteration and simulate land utilization changes.

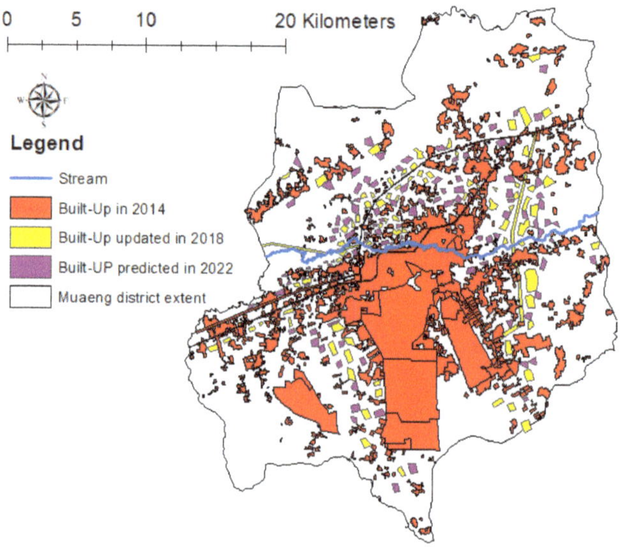


**Figure 3.** Built-up areas in 2014 and 2018 and built-up growth in 2022.

The calculation of a matrix of the CA-Markov model is different in space. This research configured the change matrix (all results of the transition matrix are shown in Tables A1–A10.) and the range of satellite imagery during the year 2014/2016/2018. Selecting a range of images used to create a transition matrix is a 2-year-old pitch, as it requires satellite imagery to show the expansion pattern of a uniform building, and the buildings are scattered with similar shapes and orientations. The model can be used to predict the position of the buildings that will be built in future years according to the assumptions of the duration of the satellite imagery range. In other words, the range of images must not be too sparse, and the relation to the land-use must have a similar pattern. However, with the ability of the model to allow the modification of the iteration loop, this allows the model to predict the expansion space of the buildings for more than 2 years. However, this research focuses on increasing the loop so that the model is able to predict future built-up areas to a satisfactory $R^2$ level. Table 1 displays the result of the matrix result changes. The comparison of land-use classification with visual interpretation data is shown in Tables A2–A4. From Appendix A the result of the error matrix for accuracy assessment of land-use types in 2014 found that the overall accuracy = 100% and with Kappa coefficient of agreement = 1.0.

**Table 1.** Result comparison of built-up growth simulation using the CA-Markov model with visual interpretation.

| Iterations-Loop | ($R^2$) from Image Simulated Using Transition Matrix 2014/2016 with Image Digitized in 2018 | ($R^2$) from Image Simulated Using Transition Matrix 2016/2018 with Image Digitized in 2018 |
|---|---|---|
| 1 | 0.91 | 0.95 |
| 2 | 0.87 | 0.92 |
| 3 | 0.85 | 0.91 |
| 4 | 0.81 | 0.86 |
| 5 | 0.77 | 0.79 |

In addition to the error matrix for accuracy assessment inspection in this study, it is used to create buffer areas out of the predict built-up to compare them with built-up from visual interpretation. The size of these two building areas is superimposed with the Overlay process.

Recursive loop assignments are defined only around the 1–5 range for a matrix to change the pitch of a 2-year interval to be used to synthesize the most appropriate iteration loop that is able to predict the expanding position of the built-up area in the future year, and also the precise test result of the model. Comparison of the positioning accuracy of the simulated image from the model to the size of the closed-space image translates from the satellite to the extent that the allowable configuration of the surplus boundary must be less than 0.04 sq. km, which indicates that the polygon is the same position.

The analysis result of Table 1 showed that the model could create a transition matrix from the years 2014–2016 to predict the position of the built-up polygon in the year 2018 and found that the $R^2$ is greater than the level 0.81 from the determination of iterations from the range of 1–4. The precision trend is gradually reduced when a loop is set to cycle 5. The accuracy trend is reduced gradually when a loop is set to five rounds and this indicates that the ability of the loop to be determined should not exceed four rounds to be used for future built-up position forecasts. Therefore, this research was made to predict the expansion position of the built-up area in the year 2022, which was constructed from the 2018 image range matrix, combined with four rounds of iterations. It is confirmed that the CA model can predict the buildings in both error matrix and buffer zone tests, with the confidence level of decision-making more than 81%.

The rise of built-up forecasts found that the area near the end of the Motorway-6 (yellow polyline) within a radius of 2 km has a higher growth rate than other areas. Although this area has a lower elevation than the other zone areas, it is also near the Lamtaklong River, as shown in Figure 3. Changes to the location of the built-up areas are mainly changes to the miscellaneous area. The water and road areas are rarely altered, although when more iterations are prescribed, the growth rate of the built-up area is higher, but the growth rate is relatively fixed within a radius of 5 km around the river. The result of the built-up area's expansion forecast in the year 2022 found that the built-up area had expanded along both sites of the road, because this zone area has a dense number of buildings and was evenly dispersed before the construction of the Highway was approved. The existing built-up area influences the model to calculate the transition matrix value, making the forecast with a very high-definition a highly-reproducible cycle, often with high accuracy. A cluster of existing buildings can also be attached, allowing the open space to be transformed into a built-up area with the driving force of economic and social factors, making fast changes to land utilization.

The application of a CA-Markov model in predicting the growth of most buildings considers the following points. The integration of the CA-Markov model is considered to be valuable for modelling land-use changes and is able to simulate and predict changes [42,43]. The CA-Markov model is the combination of Cellular Automata and transition probability matrix generated by the cross tabulation of two different images [42]. This combination of the CA-Markov model provides a robust approach in spatio-temporal dynamic modelling [42,44]. Furthermore, CA uses Markov to add spatial character to the model. On other words, the CA-Markov chain can simulate two-way transitions among any number of categories and can predict any transition among any number of categories [45,46]. It is worth mentioning that, the Cellular Automata is a dynamic process model that is used for the land-use cover change. This kind of model is common in the land-use modelling literature. Each cell with their own characteristics can represent parcels of land and can represent self-growth interactions as they are dynamic and reduplicate [47]. Furthermore, the land-use changes for any location (cells) can be clarified by the existing state and changes in neighbouring cells and can simulate the growth of things in two directions. This model is broadly used in spatial model for predicting future land-use [43,46].

The precision of the CA-Markov model needs to be monitored in use, but in this study it was confirmed that the model can be predicted at a satisfactory significant level. The CA-Markov model can predict the borderless growth of built-up in the Korat area. However, the change in land-use is a force that promotes built-up with only an expanding area zone, requiring high spatial resolution data to be used to construct replicas. The resolution of spatial data used in this study confirms that it is appropriate and directly affects the accuracy of the model. Sometimes, the precision of the model deviates with the position of the point in comparison with the result of the model. The accuracy test in this study was used to randomly sample points of each type of land-use. Dividing a basin into a sub-size allows the CA-Markov model to generate a number of data to calculate the probability more precisely. Using a 30 m spatial resolution data can create a boundary that can lead to an independent variable value.

### 3.2. DEM Change and Hydrological Characteristics in 2022

The result of DEM modeling was used to predict the future of numerical height with a linear regression model. The data used for modeling is obtained from the 58 field observation points as a survey by setting the model's altitude measuring surveyor's telescope. The original altitude of the 2014 base year was from the Royal Thai Survey Department (RTSD) under the Ministry of Defense (MOD), Bangkok of Thailand [37]. The altitude information used as the basis for this study was made by flying with an aircraft of the RTSD agencies, so the spatial resolution is more detailed than other types of data available. It also updates this information up to date from exploring the field's altitude. This is in addition to the use of the proximity measurement method from the angle of contact with the river line with the function (Near) of the QGIS 3.6.0 to measure the distance between the altitude and the angle of contact with the river line. These factors are taken as independent variant information in a linear model. As a result, the data used to create the model in Table 2 results in Equation (14) and creates a data layer. The height of DEM in 2022 for hydrological characteristic analysis shows the altitude change of DEM, as Figure 4a–c.

$$DEM_{n+1} = 1.02\ (DEM_n) - (0.0000927 \times L_i) \tag{14}$$

where $DEM_{n+1}$ is ($Y$, dependent variable) the elevated point at any $i$ in 2022, $DEM_n$ is ($X1$) the elevated point of cell any $i$ in 2018 (meters) and $L_i$ is ($X2$) the distance between the center point of built-up polygons of feature any $i$ to the closet Lamtaklong stream line feature $s$ (meters), as shown in the data input of the model in Table 2. The DEM data is re-adjusted with the Fill tool and Sink to smooth surface adjustment and water flow.

A model with an $R^2$ value of 0.84 variable coefficient ($X2$) has a negative relationship with the altitude value. The utilization form of land buildings near the river is less than 2 km, and there is a higher filling rate than the areas further away. This linear model confirms that the built-up area is far away from the river, Most of these areas are the height of the area, less than the area near the river. Some areas have an increase in altitude, as these areas comprise the built-up type of housing, which must be filled at a height of more than 4 m.

The variable that influences altitude the most is DEM ($X1$). The elevation value of the $n$th base-year that is known to be influenced by the 1.02 model is when other independent variables have a value of 0. This linear regression model is used to predict the future height in the same position found to change the utilization of land to the built-up area. This cell is recalculated to the new level (1.02 × 180 = 183.60 m) which causes the altitude value of the elevation to be changed. The DEM model predicts the elevation map in 2022 from the use of image maps of built-up areas in 2022, imported, and processed in conjunction with the DEM-year-2018 old map, which has a height point updated from field exploration. This is shown in Table 2 and Figure 4.

**Table 2.** Data for DEM modeling.

| No. | Y (m) | $X_1$ (m) | $X_2$ (m) | Predicted (m) | Residual (m) | No. | Y (m) | $X_1$ (m) | $X_2$ (m) | Predicted (m) | Residual (m) |
|---|---|---|---|---|---|---|---|---|---|---|---|
| 1 | 165 | 163 | 3700 | 165.89 | 0.89 | 30 | 181 | 179 | 1300 | 180.38 | −0.62 |
| 2 | 167 | 164 | 3600 | 168.05 | 1.05 | 31 | 180 | 177 | 1100 | 179.22 | −0.78 |
| 3 | 172 | 167 | 2550 | 172.33 | 0.33 | 32 | 174 | 172 | 1800 | 173.99 | −0.01 |
| 4 | 180 | 175 | 1220 | 180.49 | 0.49 | 33 | 172 | 171 | 4000 | 172.38 | −0.38 |
| 5 | 187 | 183 | 1201 | 186.83 | −0.17 | 34 | 180 | 179 | 1500 | 180.54 | 0.54 |
| 6 | 165 | 160 | 3405 | 164.63 | −0.37 | 35 | 187 | 184 | 900 | 185.82 | −1.18 |
| 7 | 192 | 187 | 550 | 191.99 | −0.01 | 36 | 165 | 163 | 4200 | 163.65 | −1.35 |
| 8 | 176 | 173 | 4450 | 176.58 | 0.58 | 37 | 192 | 188 | 600 | 191.01 | −0.99 |
| 9 | 188 | 184 | 375 | 186.23 | −1.77 | 38 | 176 | 173 | 4200 | 176.55 | 0.55 |
| 10 | 179 | 176 | 2500 | 178.50 | −0.50 | 39 | 188 | 185 | 980 | 187.27 | −0.73 |
| 11 | 186 | 183 | 1112 | 184.49 | −1.51 | 40 | 179 | 174 | 1700 | 177.53 | −1.47 |
| 12 | 185 | 181 | 800 | 184.36 | −0.64 | 41 | 186 | 184 | 960 | 187.50 | 1.50 |
| 13 | 182 | 176 | 1050 | 180.40 | −1.60 | 42 | 185 | 184 | 1300 | 184.41 | −0.59 |
| 14 | 181 | 177 | 900 | 180.40 | −0.60 | 43 | 181 | 179 | 1200 | 180.38 | −0.62 |
| 15 | 180 | 177 | 920 | 178.05 | −1.95 | 44 | 180 | 177 | 1100 | 179.12 | −0.88 |
| 16 | 174 | 170 | 3200 | 174.23 | 0.23 | 45 | 174 | 171 | 2560 | 174.15 | 0.15 |
| 17 | 177 | 173 | 1200 | 177.16 | 0.16 | 46 | 177 | 173 | 2700 | 178.27 | 1.27 |
| 18 | 179 | 176 | 800 | 180.39 | 1.39 | 47 | 179 | 176 | 2100 | 179.41 | 0.41 |
| 19 | 182 | 178 | 920 | 182.49 | 0.49 | 48 | 182 | 180 | 1400 | 183.55 | 1.55 |
| 20 | 187 | 181 | 945 | 186.43 | −0.57 | 49 | 187 | 185 | 960 | 186.38 | −0.62 |
| 21 | 180 | 179 | 500 | 181.45 | 1.45 | 50 | 180 | 178 | 1300 | 179.35 | −0.65 |
| 22 | 181 | 180 | 3600 | 180.65 | −0.35 | 51 | 181 | 176 | 1250 | 182.65 | 1.65 |
| 23 | 192 | 188 | 1300 | 191.09 | −0.91 | 52 | 192 | 189 | 700 | 191.12 | −0.88 |

**Table 2.** *Cont.*

| No. | Y (m) | $X_1$ (m) | $X_2$ (m) | Predicted (m) | Residual (m) | No. | Y (m) | $X_1$ (m) | $X_2$ (m) | Predicted (m) | Residual (m) |
|---|---|---|---|---|---|---|---|---|---|---|---|
| 24 | 176 | 174 | 1700 | 174.52 | −1.48 | 53 | 176 | 174 | 3200 | 177.57 | 1.57 |
| 25 | 188 | 185 | 1300 | 186.31 | −1.69 | 54 | 188 | 186 | 980 | 188.33 | 0.33 |
| 26 | 179 | 176 | 1700 | 177.31 | −1.69 | 55 | 179 | 177 | 1700 | 179.30 | 0.30 |
| 27 | 186 | 174 | 1300 | 184.45 | −1.55 | 56 | 180 | 177 | 1900 | 180.35 | 0.35 |
| 28 | 185 | 181 | 1250 | 183.38 | −1.62 | 57 | 181 | 177 | 1400 | 190.95 | −0.05 |
| 29 | 182 | 177 | 1100 | 182.40 | −0.40 | 58 | 175 | 171 | 4520 | 174.40 | −0.60 |

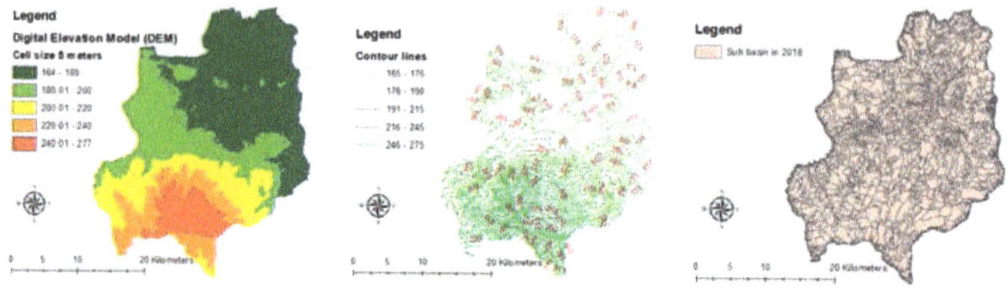


(a) reclassification of DEM (5 m) in 2022

(b) Contour line interval 5 m

(c) Sub-basin

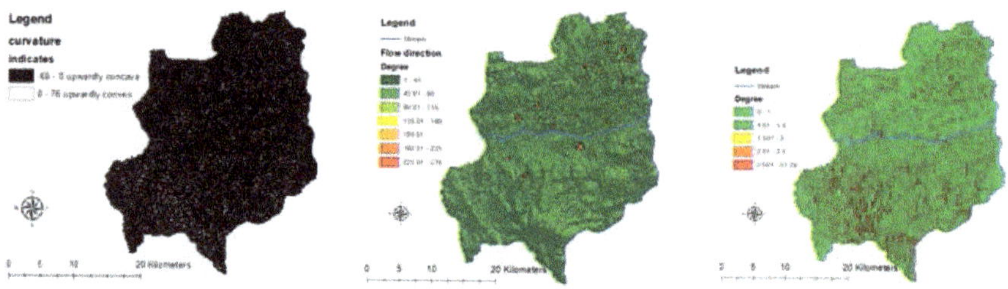


(d) Curvature

(e) Flow direction

(f) Slope

**Figure 4.** Maps of hydrological data (**a–f**) to use for independent variables extraction of FRS-GWR model in 2022.

From Figure 5a–c, the average altitude is seen to have significantly changed in the upper area of the Lamtaklong River mainly, and the result of significant hydrologic change in Table 3, the number of sub-basins generated from the data DEM Years 2014, 2018, and 2022, found that there was a total number of 2325, 2151, 2050 polygons, respectively. The number of sub-basins tends to be less likely to decline in future years as they increase the level of numerical height in the joint's area of the sub-basin region. The area of any adjacent sub-basin is fused together and water can flow to each area within the basin. This results in a longer flow sequence length from 878.47, 888.58, and 890.32 km in 2014, 2018, and 2022 respectively. There are several simple studies that say that DEM is converted to affect the nature of the basin area. The definitions of drainage and relief are essential for understanding spatial differences within the catchment [48]. Drainage density has been found to correlate with valley density, channel head source area, relief, climate, vegetation, soil and rock properties, and landscape evolution processes [49]. Measuring drainage density is extremely difficult, and it relies on good topographic maps at a detailed scale [50,51]. As an alternative to drainage density, the parameter of potential drainage density is often obtained from a digital elevation model (DEM) [52]. Drainage density has been found to correlate with valley density, channel head source area, relief, climate, vegetation, soil and rock properties, and landscape evolution processes [49]. Analysis by Pal and Saha [48] showed a high correlation between drainage density and the following parameters: length of overland flow, number of stream junctions in the basin, and the infiltration coefficient and drainage texture [48]. Low drainage diversity was related to low drainage density. A guideline for modeling of this research synthesized variables that influence the changing of the height in a physical area into an independent variant, which allows the modeling to predict the hydrologic change.

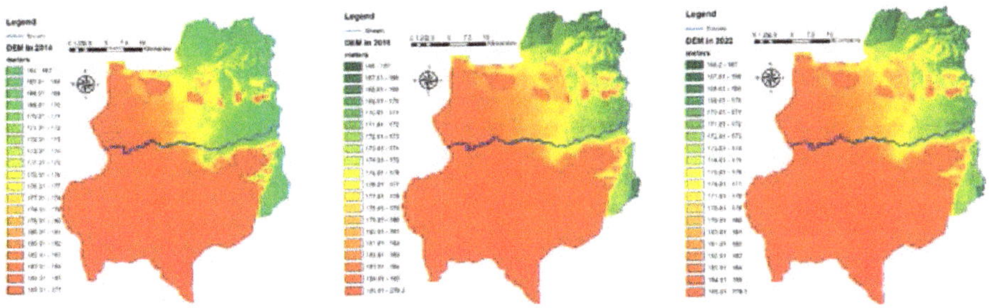

(**a**) DEM in 2014 (**b**) DEM in 2018 (**c**) DEM in 2022

**Figure 5.** Maps of DEM in (**a**) Year 2014, (**b**) Year 2018 and (**c**) Year 2022.

**Table 3.** DEM change impact to basic hydrological characteristics.

| DEM of Year | Number of Sub-Basins | Number of Sub-Basins in Upper Streams | Length of Stream Order 3 to 6 (Kilometers) | Average Elevation Range (165–175 m) in Upper Streams |
|---|---|---|---|---|
| 2014 | 2325 | 1185 | 878.47 | 173.54 |
| 2018 | 2151 | 1078 | 888.58 | 175.45 |
| 2022 | 2050 | 984 | 890.32 | 177.58 |

The hydrologic base data that is used to calculate the index of independent variables of the year 2022. Figure 4a is a 5 m data class DEM, a resolution of 5 m, which is divided into a range of 180 over 20 m in height. The DEM 2022 data is used as the basis for the creation of all other information; (b) contour line interval 5 m; (c) any sub-basin; (d) surface curvature; (e) flow direction; and (f) slope.

The entire data layer uses the edited data modification process (manipulation using GIS process) to be able to analyze the Geographic Information System (GIS) by assigning the cell sizes of the raster data to 5 m. The data processing takes a long time, but the results show the resolution and amount of descriptive data that will improve the accuracy of the GWR model.

The map in Figure 6 displays that the stream order is used to level 3 to 6, because it is a level with the embodiment of the flow outlet with the flow point from high to low. The stream order generated from DEM in 2018 compared to the year 2022 found that the stream order in 2022 is longer than that of the year of 2018 in every buffer distance. The rise of the building position affects the direction of the water flow. This is observed by comparing the stream length and stream order along the buffer distance every 1000 m out of the Lamtaklong River, as shown in Figure 7 and Table 4.

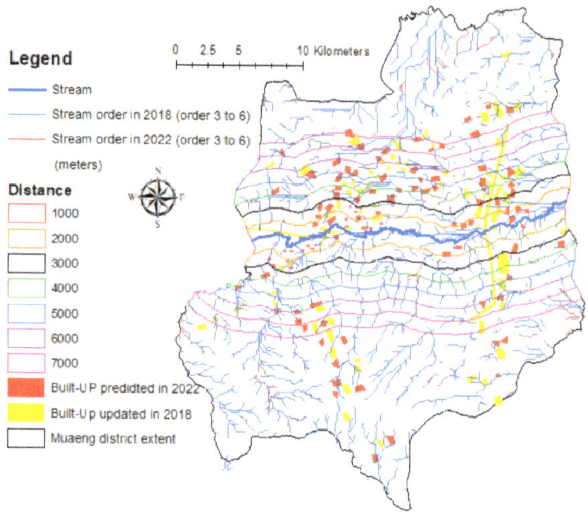


**Figure 6.** Map of stream order 3 to 6 with buffering from Lamtaklong sreamline.

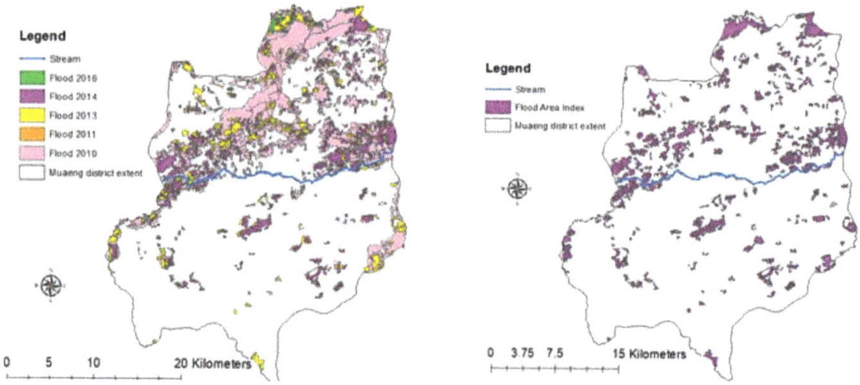


(**a**) Flooding event during 2010–2016 (**b**) Flooding frequency during 2010–2016

**Figure 7.** Maps of flood change in (**a**) and Flood Area Index (*FAI*) in (**b**).

**Table 4.** Comparison of stream order length of Year 2018/2022.

| Distance from Lamtaklong Stream (Meters) | Stream Order in 2018 | Stream Order in 2022 |
|---|---|---|
| | Order 3 to 6 (Length KM.) | Order 3 to 6 (Length KM.) |
| 1000 | 56.7 | 56.5 |
| 2000 | 63.9 | 71.2 |
| 3000 | 73.6 | 75.4 |
| 4000 | 85.4 | 86.7 |
| 5000 | 86.1 | 87.3 |
| 6000 | 87.4 | 88.7 |
| 7000 | 87.6 | 88.9 |

The longer stream length takes much longer to drain. In a buffer space of up to 1000 m, most of them have a slight slope, allowing the analysis of the length of the flow sequence to be less than the buffer range area away from the river. Also, in the buffer range, 2000–3000 m toward the west is found to have a very dense flow sequence of water.

The area has many large road construction projects. This causes changes to the route, water flow, and creates a lack of flow lines. This area is frequently affected by flooding. Data from the DEM model contributes to the *FRS-GWR* modeling support, as well as the discussion of the results.

### 3.3. Flood Area Index (FAI)

The FAI index shows the severity of the area where flooding is most frequent from the use of flood information over a 5-year period. The retrospective of the flood area in the distance edge of about 5 km from the Lamtaklong stream line is where the FAI index is taken. The link is the spatial database (Attribute) of a variable by (*Y*) to the data layer. The sub-boundary of the year 2018 is used to create the GWR model together with other independent variables.

The Dependent variables in this study were used to create an FAI index to find the frequency of floods in the area with the lead data layer. The flood boundaries are converted into binary variables (binary). If the area that has flooded is set to 1, and otherwise is 0, the analysis process requires the conversion of the vector layer (.SHP) to raster (.TIFF) and is used in overlay function with the binary model using the raster calculation function in QGIS version 3.6.0.

The high values of the flood area index are scattered in the upper area of the study area (depicted in Figure 7) and the position of the flood reaches an average altitude of about 164–167 m, while the flood area has the characteristics of land-use as agriculture; this area has also experienced flooding from the city center, making the flood boundary overlap with other areas. The average of DEM in the range of 165 to 175 m is consistent with the analysis results from FAR, thus providing a high index in the same area.

### 3.4. GWR Model and Flood Risk Susceptible (FRS) in 2022

Table 5 summarizes the results of the Global and GWR models of study areas in the Muaeng district of Nakhon Ratchasima Province. The Monte Carlo Test summary table and the GWR model calibration found that five out of the eight significant independent variables show spatial non- stationarity. In addition, the GWR model has an $R^2$ level precision at 85%, which is greater than the global model ($R^2 = 0.57$), the GWR replica creates a layer of free variable data in the GIS data layer with the interpolation method to show the spatial variation, as shown as the map in Figure 8b–f. An $R^2$ value of GWR local operation which is significant to the forecast has a value in the range of 0.8–0.88 and is an average, thus giving 0.6, and the global result of the $R^2$ value is 0.57. It can be assumed that the relationship between the selected factors and in the created city will be captured by the GWR model in those regions [41]. The growth of built-up area use in the region with low $R^2$ may be affected to a greater degree by other factors that were not considered in this study, and there may also be fringe effects that were also not considered. In the local operation of GWR, an F-test was also used to test whether spatial changes exist in the relationship under the study [41], specifically testing whether the GWR model is updated and explains significantly the relationship over the general global operation using Ordinary Least Square (OLS).

**Table 5.** Summary results of the flood risk susceptibility FRS-GWR model.

| Flood Risk Susceptibility GWR Model Coefficients | | | |
|---|---|---|---|
| | GLOBAL GWR | | |
| **Variables** | $\beta$ | $t$ | $p$**-Value** [a] |
| Intercept | 40.56 | 4.16 *** | 0.00 *** |
| Normalized Digital Elevation Model Index (*NDEMI, X1*) | −26.03 | −1.61 *** | 0.01 *** |
| Normalized Contour Index (*CLI, X2*) | −21.00 | −3.87 *** | 0.84 *n/s* |
| Area of Built-up Index (*ABUI, X3*) | 1.35 | 5.25 *** | 0.00 *** |
| Density of Built-up ($D_j$, *X4*) | 4.82 | 6.58 *** | 0.00 *** |
| Curvature Index (*CI, X5*) | −0.57 | −3.10 *** | 0.00 *** |
| Slope length Index (*SLI, X6*) | −0.48 | −2.17 *** | 0.00 *** |
| Perimeters ($P_j$, *X7*) | 0.03 | 0.29 *n/s* | 0.95 *n/s* |
| Sub-basin Area ($S_j$, *X8*) | 0.05 | 0.38 *n/s* | 0.99 *n/s* |
| *N* | 2151 | | |
| Adjusted $R^2$ | 0.57 | | 0.88 |

*** = significant at 1% level. *n/s* = not significant. [a] Results of Monte Carlo test for spatial non-stationarity [43,44].

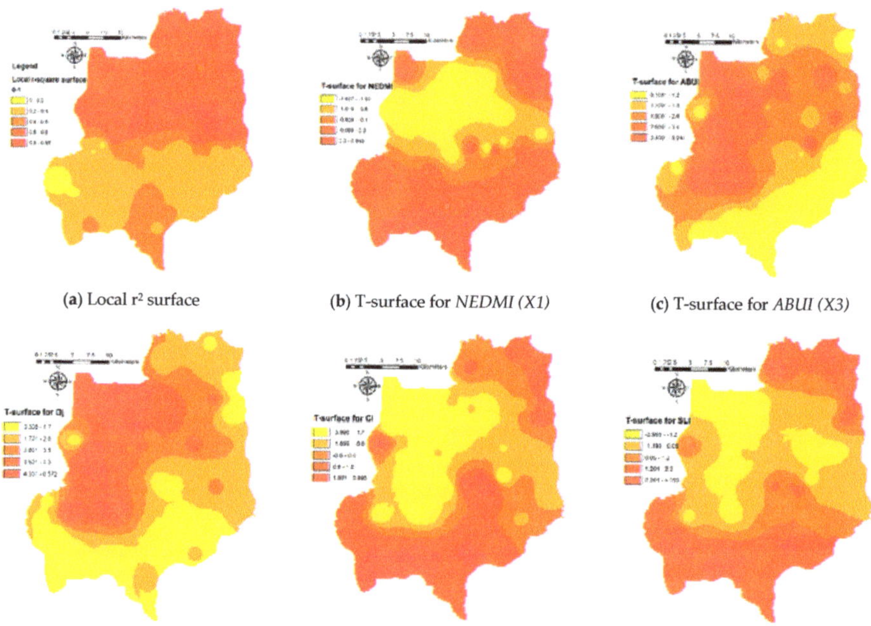

(**a**) Local r² surface (**b**) T-surface for *NEDMI (X1)* (**c**) T-surface for *ABUI (X3)*

(**d**) T-surface for $D_j$ *(X4)* (**e**) T-surface for *CI (X5)* (**f**) T-surface for *SLI (X6)*

**Figure 8.** Maps of Local r-square (**a**) and T surface distribution of FRS-GWR model (**b**–**f**).

The analysis of variance was addressed through the testing of ANOVA for the creation in the city of Korat. The F-value was 8.231. The high F-value suggests that the GWR format was significantly improved through a global form to define the relationship between the built-up growth and different factors. Additionally, the Akaike information (AIC) of the GWR format (178,474.5) is far less than one of the global operations (20,371.1). This indicates that the GWR local operation is improved more than the OLS model (referenced in Table 6).

**Table 6.** ANOVA test of the FRS-GWR over the OLS regression model.

| Source | SS | DF | MS | F |
|---|---|---|---|---|
| OLS Residuals | 56,312.3 | 26.0 | | |
| GWR Improvement | 27,006.7 | 465.36 | 4.72 | |
| GWR Residuals | 29,305.6 | 3210.53 | 7.84 | 8.231 |
| GWR Akaike Information Criterion | 178,474.5 | (OLS): 20,371.1 | | |

The *FRS-GWR* model predicts the area risk of flood sensitivity in the year 2022. The significant independent variables were only selected containing the normalized digital elevation model Index (*NDEMI*, *X*1), area of built-up index (*ABUI*, *X*3), density of built-up ($D_j$, *X*4), curvature index (*CI*, *X*5), and slope length Index (*SLI*, *X*6) and to create a model *FRS-GWR* as Equation (15) when assigning *Y* values from GWR model = $F_j$ in Equation (5). The result of the dispersion of independent variables is as follows Figure 8b–f.

$$FRS_{2022} = (4.16 - (1.61NEDMI) + (5.25ABUI) + \left(6.58D_j\right) - (3.1CI) - (2.17SLI))x100/S_j \quad (15)$$

The independent variables are selected from the relationship with the variable as a significant factor. The surface distribution coefficient (T-surface) (depicted in Figure 8) that is consistent in the independent variable area *NEDMI (X1)* shows the dispersion of negative range coefficients; in addition, however, the higher the $R^2$ high value, the higher the flooding risk. In accordance with the effect of *CI (X5)* and *SLI (X6)*, but contrary to the positive coefficient of *ABUI (X3)* and $D_j$ (*X*4,) is when there is space and the density of more built-up areas will hinder the flow of water. Additionally, the variable that is not mentioned does not have a compliance with *Y*.

The risk area forecast for flood sensitivity *FRS-GWR* model in 2022 uses the filtered independent variant data that is significant and influences the variables followed by the GIS tools simulated from a hydrologic analysis and then imported into Equation (15). Figure 9 displays the flood risk susceptibility values that range from 0.6 to 0.86, and this highlights the potential for flooding in the area when it rains. For approximately 60–86 percent of the space, most of the area is more than 70 percent of the upper basin above the Lamtaklong River. There are index values ranging from 50, but there are some areas with a lower flooding potential.

For areas with DEM ranging from 180 to 200 m in the lower part of the sub-basin, most of which is less prone to flooding as the area has an average altitude of more than the other zones, except the Eastern zone area, the index value is 40–45 because it is close to the Lamtaklong River. As well as in the year 2022, expanding the construction of the buildings in this area is rapidly emerging from the new road-building influence in the year 2018.

An *FRS-GWR* model compared to (FAI) was found to be consistent, but the *FRS-GWR* map has a more continuous dispersion of the area than the *FRS-GWR* model can predict. The sensitive area in terms of the risk of flooding and the ideal analysis in the area is not very large, and there is a large number of sub-areas (spatial units) and the continuous dispersion will greatly improve the forecast performance.

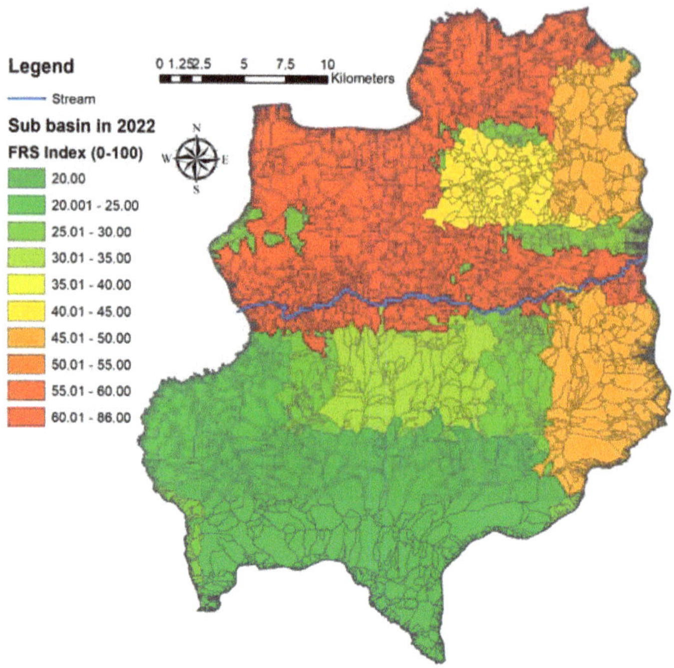


**Figure 9.** Map of flood risk susceptible forecast using *FRS*-GWR model in 2022.

## 4. Conclusions

A future building forecast with a CA-Markov model in this study found that it can be adjusted to require repeated process assignments (Iterations), making it possible to use images to predict future buildings for many years to come. This study uses land utilization data every 2 years to predict the position of buildings that are arising up to 4 years, using 2014/2016 data to predict the building of the year 2018. ($R^2$ value is also greater than 0.8). In general, predictive research with a CA-Markov model can be used to create a map based on data spacing. According to Bertaud [53], it is important to note that the areas that are configured in this way require a dispersion of the land cover that requires continuous dispersion.

This research also found that the GWR modeling provides precise prediction capabilities. It is necessary to create a layer of spatial units to be appropriate for use. Reference by Brunsdon [54], explains how the unit of this sub-area is different depending on the spatial relationship of independent variables and the variables based on this research. This research hypothesized that the data layer DEM with a higher average level will affect other hydrologic data changes and cause the sub-watershed boundary (sub-basin) to change. In order to be used to extract variable data, as well as independent variables, to be consistent and relevant in the testing area, various statistical values, including local $R^2$, AIC, and F-test, were used to confirm that the GWR model is more accurate than the GLOBAL model (OLS) in every test. The research developed the GWR model in conjunction with a simulated flood frequency index model to analyze risk areas of flood sensitivity (*FRS-GWR*) as confirmed by a model that can predict the sub-basin that has potential to be at risk of flooding in future years from a comparative test against flood data from the past.

Modeling guidelines for the relationship of spatial heights affect the analysis of water flow in both the basin region, flow accumulation, flow direction and stream order. Reference by Jenkins [29], stated that if there are many adjacent built-ups, there will result a change from an ever-changing area to a water-inlet area because the water has no outlet.

However, further research in other areas may be added to the use of TIN data (Tri Irregular Network) in conjunction with the analysis of the flow of water to provide a more complete flow direction, as well as analysis of the accumulation of water (flow accumulation). Analysis of flow accumulation enables a more continuous flow of water to accumulate and this can be used to drain the water out of the sub-basin area. However, analysis with this information is only appropriate to work in small spaces, as this takes longer to process than the data DEM.

This study can be used as a prototype to analyze environmental impacts, especially the flow and barriers of water flow from the further expansions of buildings in other areas. Reference [40] explains the modeling research in the future to assess the flood-prone area to be included in plans in order to better cope with climate change that affects the frequency and risk of flood disasters. In addition, this research will allow stakeholders to better plan their development of urban areas so as to minimize environmental impacts and make plans for the sustainable growth of a city.

**Author Contributions:** Conceptualization, P.L. and N.B.; Formal analysis, P.L. and N.B.; Investigation, N.B.; Methodology, P.L.; Project P.L and N.B.; Writing—original draft, P.L. and N.B.; and Writing—review and editing, P.L. and N.B.

**Funding:** This research project is financially supported by Faculty of Informatics, Mahasarakham University (Fast Track 2019) and Research unit of Geo-informatics for Local Development and Climate Changes, Mitigation and Adaptation Research Unit; CMARE).

**Conflicts of Interest:** The authors declare no conflict of interest.

## Appendix A

**Table A1.** Allocation for land-use categories in 2014, 2016, and 2018.

| Land-Use Types | 2014 | | 2016 | | 2018 | |
|---|---|---|---|---|---|---|
| | sq. km | % | sq. km | % | sq. km | % |
| Urban and built-up area (U) | 202.68 | 31.40 | 209.45 | 32.43 | 225.88 | 34.97 |
| Agriculture land (A) | 385.01 | 59.61 | 382.05 | 59.15 | 370.06 | 57.30 |
| Forest land (F) | 7.12 | 1.10 | 7.03 | 1.08 | 6.52 | 1.01 |
| Water body (W) | 7.45 | 1.15 | 7.06 | 1.09 | 6.91 | 1.07 |
| Miscellaneous land (M) | 43.58 | 3.33 | 40.25 | 6.25 | 36.47 | 5.65 |
| Total | 645.84 | 100.00 | 645.84 | 100.00 | 645.84 | 100.00 |

**Table A2.** Error matrix for accuracy assessment of land-use types in 2014.

| Land-Use Types in 2014 | Reference Data in 2018 | | | | | |
|---|---|---|---|---|---|---|
| | U | A | F | W | M | Total |
| Urban and built-up area (U) | 43 | 0 | 0 | 0 | 0 | 43 |
| Agriculture land (A) | 0 | 58 | 0 | 0 | 0 | 58 |
| Forest land (F) | 0 | 0 | 3 | 0 | 0 | 3 |
| Water body (W) | 0 | 0 | 0 | 15 | 0 | 15 |
| Miscellaneous land (M) | 0 | 0 | 0 | 0 | 18 | 18 |
| Total | 43 | 58 | 3 | 15 | 18 | 137 |

Note: 1. Overall accuracy = 100 %. 2. Kappa hat coefficient of agreement = 1.0.

Table A3. Error matrix for accuracy assessment of land-use types in 2016.

| Land-Use Types in 2016 | Reference Data in 2018 | | | | | |
|---|---|---|---|---|---|---|
| | U | A | F | W | M | Total |
| Urban and built-up area (U) | 57 | 0 | 0 | 0 | 0 | 57 |
| Agriculture land (A) | 0 | 60 | 0 | 0 | 0 | 60 |
| Forest land (F) | 0 | 0 | 2 | 0 | 0 | 2 |
| Water body (W) | 0 | 0 | 0 | 14 | 0 | 14 |
| Miscellaneous land (M) | 0 | 0 | 0 | 0 | 16 | 16 |
| Total | 57 | 60 | 2 | 14 | 16 | 149 |

Note: 1. Overall accuracy = 100 %. 2. Kappa hat coefficient of agreement = 1.0.

Table A4. Error matrix for accuracy assessment of land-use types in 2018.

| Land-Use Types in 2018 | Reference Data in 2018 | | | | | |
|---|---|---|---|---|---|---|
| | U | A | F | W | M | Total |
| Urban and built-up area (U) | 65 | 0 | 0 | 0 | 0 | 65 |
| Agriculture land (A) | 0 | 63 | 0 | 0 | 0 | 63 |
| Forest land (F) | 0 | 0 | 2 | 0 | 0 | 2 |
| Water body (W) | 0 | 0 | 0 | 12 | 0 | 12 |
| Miscellaneous land (M) | 0 | 0 | 0 | 0 | 14 | 14 |
| Total | 65 | 63 | 2 | 12 | 14 | 156 |

Note: 1. Overall accuracy = 100 %. 2. Kappa hat coefficient of agreement = 1.0.

Table A5. Transition area matrix for land-use change between 2014 and 2016.

| Land-Use in 2016 | Land-Use Types (sq. km) | | | | | |
|---|---|---|---|---|---|---|
| | U | A | F | W | M | Total |
| Urban and built-up area (U) | 209.45 | 0.00 | 0.00 | 0.00 | 0.00 | 209.45 |
| Agriculture land (A) | 31.52 | 330.75 | 0.00 | 1.15 | 18.63 | 382.05 |
| Forest land (F) | 0.05 | 0.62 | 5.21 | 0.01 | 1.14 | 7.03 |
| Water body (W) | 0.03 | 0.56 | 0.00 | 6.01 | 0.46 | 7.06 |
| Miscellaneous land (M) | 15.34 | 1.02 | 0.00 | 0.21 | 23.68 | 40.25 |
| Total | 256.39 | 332.95 | 5.21 | 7.38 | 43.91 | 645.84 |

Table A6. Transition probability matrix for land-use change between 2014 and 2016.

| Land-Use in 2016 | Land-Use Types | | | | | |
|---|---|---|---|---|---|---|
| | U | A | F | W | M | Total |
| Urban and built-up area (U) | 1.000 | 0.000 | 0.000 | 0.000 | 0.000 | 1.000 |
| Agriculture land (A) | 0.083 | 0.866 | 0.000 | 0.003 | 0.048 | 1.000 |
| Forest land (F) | 0.008 | 0.088 | 0.741 | 0.001 | 0.162 | 1.000 |
| Water body (W) | 0.004 | 0.079 | 0.000 | 0.852 | 0.065 | 1.000 |
| Miscellaneous land (M) | 0.381 | 0.025 | 0.000 | 0.005 | 0.589 | 1.000 |

**Table A7.** Transition area matrix for land-use change between 2016 and 2018.

| Land-Use in 2018 | Land-Use Types (sq. km) | | | | | |
|---|---|---|---|---|---|---|
| | U | A | F | W | M | Total |
| Urban and built-up area (U) | 225.88 | 0.00 | 0.00 | 0.00 | 0.00 | 225.88 |
| Agriculture land (A) | 35.61 | 316.87 | 0.00 | 1.02 | 16.56 | 370.06 |
| Forest land (F) | 0.00 | 0.23 | 5.15 | 0.02 | 1.12 | 6.52 |
| Water body (W) | 1.23 | 0.03 | 0.00 | 4.01 | 1.64 | 6.91 |
| Miscellaneous land (M) | 14.56 | 0.95 | 0.00 | 0.14 | 20.82 | 36.47 |
| Total | 277.28 | 318.08 | 5.15 | 5.19 | 40.14 | 645.84 |

**Table A8.** Transition probability matrix for land-use change between 2016 and 2018.

| Land-Use in 2018 | Land-Use Types | | | | | |
|---|---|---|---|---|---|---|
| | U | A | F | W | M | Total |
| Urban and built-up area (U) | 1.000 | 0.000 | 0.000 | 0.000 | 0.000 | 1.000 |
| Agriculture land (A) | 0.096 | 0.856 | 0.000 | 0.003 | 0.045 | 1.000 |
| Forest land (F) | 0.000 | 0.035 | 0.789 | 0.003 | 0.173 | 1.000 |
| Water body (W) | 0.179 | 0.004 | 0.000 | 0.580 | 0.237 | 1.000 |
| Miscellaneous land (M) | 0.399 | 0.026 | 0.000 | 0.003 | 0.572 | 1.000 |

**Table A9.** Transition area matrix for land-use change between 2018 and 2022.

| Land-Use in 2022 | Land-Use Types (sq. km) | | | | | |
|---|---|---|---|---|---|---|
| | U | A | F | W | M | Total |
| Urban and built-up area (U) | 242.85 | 0.00 | 0.00 | 0.00 | 0.00 | 242.85 |
| Agriculture land (A) | 42.62 | 300.06 | 0.00 | 1.03 | 18.60 | 362.31 |
| Forest land (F) | 0.00 | 0.01 | 6.17 | 0.02 | 0.02 | 6.22 |
| Water body (W) | 1.58 | 0.82 | 0.00 | 2.73 | 1.84 | 6.97 |
| Miscellaneous land (M) | 15.64 | 0.20 | 0.00 | 0.18 | 11.47 | 27.49 |
| Total | 302.69 | 301.09 | 6.17 | 3.96 | 31.93 | 645.84 |

Iterations loop = 4.

**Table A10.** Transition probability matrix for land-use change between 1986 and 1994.

| Land-Use in 2022 | Land-Use Types | | | | | |
|---|---|---|---|---|---|---|
| | U | A | F | W | M | Total |
| Urban and built-up area (U) | 1.000 | 0.000 | 0.000 | 0.000 | 0.000 | 1.000 |
| Agriculture land (A) | 0.117 | 0.828 | 0.000 | 0.003 | 0.052 | 1.000 |
| Forest land (F) | 0.000 | 0.002 | 0.991 | 0.003 | 0.023 | 1.000 |
| Water body (W) | 0.227 | 0.118 | 0.000 | 0.392 | 0.263 | 1.000 |
| Miscellaneous land (M) | 0.569 | 0.007 | 0.000 | 0.007 | 0.417 | 1.000 |

**Table A11.** Data and Equipment.

| Data and Equipment | Date | Number of Date (sheet) | Scale | Source/Remarks |
|---|---|---|---|---|
| 1. RS/GIS Data Types | | | | |
| 1.1. primary datasets-Satellite imagery | December 2014/2016/2018 | 750,000 sheets | 0.9 m × 0.9 m | Google Earth Pro |
| -Topographic map (5438IV) | 2018 | 1 | 1:50,000 | Royal Thai Survey Department (RTSD) |
| 1.2. Secondary datasets | | | | |
| -Land-use | 2014/2016/2018 | - | 1:4,000 | Land Development Department (LDD) and updated from field observation |
| Road | 2018 | - | - | Department of Highways (DOH) and updated from field observation |
| Stream | 2016 | - | - | Department of Water Resources |
| Contour line | - | - | - | field observation |
| DEM | 2018 | - | 5 m × 5 m | (RTSD), FGDS from Geo-Informatics and Space Technology Department Agency (GISTDA) and updated from field observation |
| 2. Equipment hardware and software | | | | |
| 2.1. GPS | | | | -Research unit of Geo-informatics for Local Development |
| 2.2. Notebook (Acer Aspire VX15) | | | | -Personal |
| 2.3 Software<br>- QGIS 3.6.0<br>- IDRISI 15.0<br>- ArcGIS 9.3.1 | | | | -Free<br>-Personal<br>-Research unit of Geo-informatics for Local Development |

**Table A12.** List of abbreviation.

| The Description | Abbreviation |
|---|---|
| Akaike Information Criterion | AIC |
| Area of Built-Up Index | (ABUI, X3) |
| Arc Geographic Information System | ArcGIS |
| Cellular Automata | (CA) |
| Conversion of land use changes and its Effects at Small regional extent | CLUE-S |
| Curvature Index | (CI, X5) |
| Degree of Freedom | DF |
| Density of Built-Up | ($D_j$, X4) |
| Digital Elevation Model | (DEM) |
| Flood Risk Susceptibility (FRS) in GWR | FRS-GWR |
| F-Ratio | F |
| Geographic Weighted Regression | (GWR) |

**Table A12.** *Cont.*

| The Description | Abbreviation |
|---|---|
| Geographic Resources Analysis Support System | GRASS |
| Integrated Geographic Information System | (GIS) |
| and remote sensing software | IDRISI |
| Land Transformation Model | LTM |
| Land use | (LU) |
| Mean square | MS |
| Nakhon Ratchasima province | some called as Korat |
| Normalized Contour Index | (CLI, X2) |
| Normalized Digital Elevation Model Index | (NDEMI, X1) |
| Ordinary Least Square | OLS |
| Quantum GIS software package | QGIS 3.6.0 |
| Raster format in IDRISI | (.RST) |
| Slope Length Index | (SLI, X6) |
| Sum of Square | SS |
| Vector format in ArcGIS | (.SHP) |

## References


1. Winston, N. From Boom to Bust? An Assessment of the Impact of Sustainable Development Policies on Housing in the Republic of Ireland. *Local Environ. Int. J. Justice Sustain.* **2007**, *12*, 57–71. [CrossRef]
2. Galster, G.; Hanson, R.; Ratcliffe, M.R.; Wolman, H.; Coleman, S.; Freihage, J. Wrestling Sprawl to the Ground: Defining and measuring an elusive concept. *Hous. Policy Debate* **2001**, *12*, 681–717. [CrossRef]
3. Frenkel, A.; Ashkenazi, M. Measuring urban sprawl: How can we deal with it? *Environ. Plan. B* **2008**, *35*, 56–79. [CrossRef]
4. Arribas-Bel, D.; Nijkamp, P.; Scholten, H. Multidimensional urban sprawl in Europe: A self-organizing mapapproach. *Comput. Environ. Urban Syst.* **2011**, *35*, 263–275. [CrossRef]
5. Hamidi, S.; Ewing, R.; Preuss, I.; Dodds, A. Measuring Sprawl and Its Impacts: An Update. *J. Plan. Educ. Res.* **2015**, *35*, 35–50. [CrossRef]
6. Department of Public Works and Town & Country Planning. Final Annual Report. Available online: thttp://pvnweb.dpt.go.th/nakhonratchasima/index.php/2018-01-10-07-18-02 (accessed on 18 February 2018).
7. Ewing, R.; Schmid, T.; Killingsworth, R.; Zlot, A.; Raudenbush, S. Relationship between Urban Sprawl and Physical Activity, Obesity, and Morbidity. *Am. J. Health Promot.* **2003**, *18*, 47–57. [CrossRef]
8. Lee, A.C.K.; Maheswaran, R. The health benefits of urban green spaces: A review of the evidence. *J. Public Health* **2011**, *33*, 212–222. [CrossRef]
9. Brueckner, J.K.; Largey, A.G. Social interaction and urban sprawl. *J. Urban Econ.* **2008**, *64*, 18–34. [CrossRef]
10. McColl, C.; Aggett, G. Land-use forecasting and hydrologic model integration for improved land-use decision support. *J. Environ. Manag.* **2007**, *84*, 494–512. [CrossRef]
11. Wang, Y.; Chen, A.S.; Fu, G.; Djordjeviç, S.; Zhang, C.; Savíc, D.A. An integrated framework for high resolution urban flood modelling considering multiple information sources and urban features. *Environ. Model. Softw.* **2018**, *107*, 85–95. [CrossRef]
12. Zhang, B.; Xie, G.; Li, N.; Wang, S. Effect of urban green space changes on the role of rainwater runoff reduction in Beijing, China. *Landsc. Urban Plan.* **2015**, *140*, 8–16. [CrossRef]
13. Kibler, D.F. Urban stormwater hydrology. *Water Resour. Monogr.* **1982**, *7*, 1–271.
14. Cheng, S.; Wang, R. An approach for evaluating the hydrological effects of urbanization and its application. *Hydrol. Process.* **2002**, *16*, 1403–1418. [CrossRef]
15. Soulis, K.X.; Dercas, N.; Papadaki, C.H. Effects of forest roads on the hydrological response of a small-scale mountain watershed in Greece. *Hydrol. Process.* **2015**, *29*, 1772–1782. [CrossRef]
16. Mitsova, D. Coupling Land Use Change Modeling with Climate Projections Catchment Near Cincinnati, Ohio. *Int. J. Geo-Inf.* **2014**, *3*, 1256–1277. [CrossRef]
17. Goudie, A. *The Human Impact on the Natural Environment: Past, Present, and Future*; John Wiley & Sons, Ltd.: San Francisco, CA, USA, 2006.

18. Weng, Q. Modeling Urban Growth Effects on Surface Runoff with the Integration of Remote Sensing and GIS. *Environ. Manag.* **2001**, *28*, 737–748. [CrossRef]
19. Sheng, J.; Wilson, J.P. Watershed urbanization and changing flood behavior across the Los Angeles metropolitan region. *Nat. Hazards* **2009**, *48*, 41–57. [CrossRef]
20. Aparicio, J.; Martínez-Austria, P.F.; Güitrón, A.; Ramírez, A.I. Floods in Tabasco, Mexico: A diagnosis and proposal for courses of action. *J. Flood Risk Manag.* **2009**, *2*, 132–138. [CrossRef]
21. Xia, J.; Qiao, Y.F.; Song, X.F.; Ye, A.Z.; Zhang, X.C. Analysis about Effect Rules of Underlying Surface Change to the Relationship between Rainfall and Runoff in the Chabagou Catchment. *Resour. Sci.* **2007**, *29*, 71–76.
22. Feng, H.; Wu, S.F.; Wu, P.T.; Li, M. Study on Scouring Experiment of Regulating Runoff in Grassland Slope. *J. Soil Water Conserv.* **2005**, *109*, 23–25.
23. Ahrens, A.; Lyons, S. Changes in Land Cover and Urban Sprawl in Ireland from a Comparative Perspective Over 1990–2012. *Land* **2019**, *8*, 16. [CrossRef]
24. Areu-Rangel, O.S.; Cea, L.; Bonasia, R.; Espinosa-Echavarria, V.J. Impact of Urban Growth and Changes in Land Use on River Flood Hazard in Villahermosa, Tabasco (Mexico). *Water* **2019**, *11*, 304. [CrossRef]
25. Zúñiga, E.; Magaña, V. Vulnerability and risk to intense rainfall in Mexico: The effect of land use cover change. *Investig. Goegr.* **2018**, *95*, 1–18.
26. Nuñez, A.J. Plan Estatal de Desarrollo 2013–2018. Available online: https://tabasco.gob.mx//sites/all/files/sites/tabasco.gob.mx/files/pled-2013-2018_web.pdf (accessed on 2 February 2019).
27. Bermúdez, M.; Neal, J.C.; Bates, P.D.; Coxon, G.; Freer, J.E.; Cea, L.; Puertas, J. Quantifying local rainfall dynamics and uncertain boundary conditions into a nested regional-local flood modeling system. *Water Resour. Res.* **2017**, *53*, 2770–2785. [CrossRef]
28. Zhang, Y.; Zhao, Y.; Wang, Q.; Wang, J.; Li, H.; Zhai, J.; Zhu, Y.; Li, J. Impact of Land Use on Frequency of Floods in Yongding River Basin, China. *Water* **2016**, *8*, 401. [CrossRef]
29. Jenkins, K.; Surminski, L.; Hall, J.; Crick, F. Assessing surface water flood risk and management strategies under future climate change: Insights from an Agent-Based model. *Sci. Total Environ.* **2017**, *595*, 159–186. [CrossRef] [PubMed]
30. Liu, H.; Wang, Y.; Zhang, C.; Chen, A.S.; Fu, G. Assessing real options in urban surface water flood risk management under climate change. *Nat. Hazards* **2018**, *94*, 1–18. [CrossRef]
31. Pontius, R.G., Jr. Quantification error versus location error in comparison of categorical maps. *Photogramm. Eng. Remote Sens.* **2000**, *66*, 1011–1016.
32. Eastman, J.R. *IDRISI Andes Guide to GIS and Image Processing*; Clark Labs, Clark University: Worcester, MA, USA, 2006.
33. Jantz, C.A.; Goetz, S.J.; Shelley, M.K. Using the SLEUTH urban growth model to simulate the impacts of future policy scenarios on urban land use in the Baltimore-Washington metropolitan area. *Environ. Plan. B Plan. Des.* **2003**, *31*, 251–271. [CrossRef]
34. Conclelis, H. From cellular automata models to urban models: New principles for model development and implementation. *Environ. Plan.* **1997**, *24*, 165–174. [CrossRef]
35. Feoli, E. *Migration Towards the Cities: Measuring the Effects of Urban Expansion in Rural-Urban Interface by GIS and RS Technology*; Department of Biology University of Trieste: Trieste, Italy, 2004.
36. Benenson, I.; Torrens, P.M. (Eds.) Introduction to Urban Geosimulation. In *Geosimulation: Automata-Based Modeling of Urban Phenomena*; John Wiley & Sons: Hoboken, NJ, USA, 2004; pp. 1–19.
37. Suwit, O. Optimum Predictive Model for Urban Growth Prediction. *Suranaree J. Sci. Technol.* **2011**, *18*, 141–152.
38. Dourvas, N.; Sirakoulis, G.; Adamatzky, A. Cellular automaton Belousov-Zhabotinsky model for binary full adder. *Int. J. Bifurc. Chaos.* **2017**, *6*, 1–15. [CrossRef]
39. Balzter, H.; Braun, P.W.; Köhler, W. Cellular Automata Models for Vegetation Dynamics. *Ecol. Model.* **1998**, *107*, 113–125. [CrossRef]
40. Fotheringham, A.S.; Brunsdon, C.; Charlton, M. *Geographically Weighted Regression: The Analysis of Spatially Varying Relationships*; John Wiley and Sons Ltd.: West Sussex, UK, 2002.
41. Noresah, M.S.; Ruslan, R. Modelling urban spatial structure using Geographically Weighted Regression. In Proceedings of the 18th World IMACS/MODSIM Congress, Cairns, Australia, 13–17 July 2009; pp. 1950–1956. Available online: http://mssanz.org.au/modsim09 (accessed on 23 March 2019).

42. Singh, S.K.; Mustak, S.; Srivastava, P.K.; Szabó, S.; Islam, T. Predicting spatial and decadal LULC changes through cellular automata Markov chain models using earth observation datasets and geo-information. *Environ. Process.* **2015**, *2*, 61–78. [CrossRef]
43. Parsa, V.A.; Yavari, A.; Nejadi, A. Spatio-temporal analysis of land use/land cover pattern changes in Arasbaran Biosphere Reserve: Iran. *Model. Earth Syst. Environ.* **2016**, *2*, 1–13. [CrossRef]
44. Wang, Y.; Zhang, X. A dynamic modeling approach to simulating socioeconomic effects on landscape changes. *Ecol. Model.* **2001**, *140*, 141–162. [CrossRef]
45. Pontius, G.R.; Malanson, J. Comparison of the structure and accuracy of two land change models. *Int. J. Geogr. Inf. Sci.* **2005**, *19*, 243–265. [CrossRef]
46. Ye, B.; Bai, Z. Simulating land use/cover changes of Nenjiang County based on CA-Markov model. *Comput. Comput. Technol. Agric.* **2008**, *1*, 321–329.
47. Brown, D.G.; Walker, R.; Manson, S.; Seto, K. Modeling land use and land cover change. In *Land Change Science*; Springer: Dordrecht, The Netherlands, 2012; pp. 395–409.
48. Pal, S.; Saha, T.K. Exploring drainage/relief-scape sub-units in Atreyee river basin of India and Bangladesh. *Spat. Inf. Res.* **2017**, *25*, 685–692. [CrossRef]
49. Rai, P.K.; Mohan, K.; Mishra, S.; Ahmad, A.; Mishra, V.N. A GIS-based approach in drainage morphometric analysis of Kanhar River Basin, India. *Appl. Water Sci.* **2017**, *7*, 217–232. [CrossRef]
50. Tucker, G.E.; Catani, F.; Rinaldo, A.; Bras, R.L. Statistical analysis of drainage density from digital terrain data. *Geomorphology* **2001**, *36*, 187–202. [CrossRef]
51. Dobos, E.; Daroussin, J. Potential drainage density Index (PDD). In *An SRTM-Based Procedure to Delineate SOTER Terrain Units on 1:1 and 1:5 Million Scales*; Office for Official Publications of the European Communities: Luxemburg, 2005; pp. 40–51.
52. Nevena, D.; Barbara, K.; Nevenka, O. Different Approaches to Estimation of Drainage Density and Their Effect on the Erosion Potential Method. *Water* **2019**, *11*, 593.
53. Bertaud, A.; Stephen, M. *The Spatial Distribution of Population in 48 World Cities: Implications for Economies in Transition*; The Centre for Urban Land Economic Research, University of Wisconsin: Madison, WI, USA, 2003; Available online: http://alain-bertaud.com/AB_Files/SpatiaDistributionofPop50Cities.pdf (accessed on 13 April 2019).
54. Brunsdon, C.; Fotheringham, A.S.; Charlton, M.E. Geographically Weighted Regression: A Method for Exploring Spatial Nonstationarity. *Geogr. Anal.* **1996**, *28*, 281–298. [CrossRef]

water

MDPI

Article

# Using High-Density LiDAR Data and 2D Streamflow Hydraulic Modeling to Improve Urban Flood Hazard Maps: A HEC-RAS Multi-Scenario Approach

**Alin Mihu-Pintilie [1,*], Cătălin Ioan Cîmpianu [2], Cristian Constantin Stoleriu [2], Martín Núñez Pérez [3] and Larisa Elena Paveluc [2,4]**

1 Institute for Interdisciplinary Research, Science Research Department, Alexandru Ioan Cuza University of Iaşi (UAIC), St. Lascăr Catargi 54, 700107 Iaşi, Romania

2 Department of Geography, Faculty of Geography and Geology, Alexandru Ioan Cuza University of Iaşi (UAIC), Bd. Carol I 20A, 700505 Iaşi, Romania

3 Escuela de Organización Industrial (EOI), Av. de Gregorio del Amo 6, 28040 Madrid, Spain

4 National Administration Romanian Waters, Siret Water Basin Administration, St. Cuza Voda 1, 600274 Bacău, Romania

* Correspondence: mihu.pintilie.alin@gmail.com; Tel.: +40-741-912-245

Received: 31 July 2019; Accepted: 1 September 2019; Published: 3 September 2019

**Abstract:** The ability to extract streamflow hydraulic settings using geoinformatic techniques, especially in high populated territories like urban and peri-urban areas, is an important aspect of any disaster management plan and flood mitigation effort. 1D and 2D hydraulic models, generated based on DEMs with high accuracy (e.g., Light Detection and Ranging (LiDAR)) and processed in geographic information systems (GIS) modeling software (e.g., HEC-RAS), can improve urban flood hazard maps. In this study, we present a small-scale conceptual approach using HEC-RAS multi-scenario methodology based on remote sensing (RS), LiDAR data, and 2D hydraulic modeling for the urban and peri-urban area of Bacău City (Bistriţa River, NE Romania). In order to test the flood mitigation capacity of Bacău 1 reservoir (rB1) and Bacău 2 reservoir (rB2), four 2D streamflow hydraulic scenarios (s1–s4) based on average discharge and calculated discharge (s1–s4) data for rB1 spillway gate (Sw1) and for its hydro-power plant (H-pp) were computed. Compared with the large-scale flood hazard data provided by regional authorities, the 2D HEC-RAS multi-scenario provided a more realistic perspective about the possible flood threats in the study area and has shown to be a valuable asset in the improvement process of the official flood hazard maps.

**Keywords:** Light Detection and Ranging (LiDAR); HEC-RAS; 2D modeling; flood hazard; urban and peri-urban area

## 1. Introduction

### 1.1. State-of-the-Art

In the last decades, with climate change and global warming, the associated natural disasters have reached more disastrous and catastrophic scales [1–6]. Many natural disaster patterns have diversified and modified under the pressure of climate change [7–9]. Among them, the intensification of the hydrological cycle has made an unprecedented impact on the magnitude, spatial extend, duration and frequency of hydro-meteorological disaster events [10–12]. Supporting this statement, many scientific publications support the fact that the occurrence of climate-related disasters (e.g., mainly floods, severe storms, cyclones, typhoons, droughts) have significantly increased under the abrupt changes in hydrological climatic conditions, the ecosystems resilience to transitional (wet–dry), and other related disturbances [13–15]. According to the Emergency Events Database (EM-DAT), in the last 50 years, climate related disasters have increased trifold between 1970 and 2017, from an average of 90 events in the period 1970–1989, to an average of more than 300 events in the period that followed after the year 2000 [16,17]. Among them, flooding phenomena are the most widespread, frequent, and costly natural disaster for the human societies [18,19], and is the most common natural hazard and the third most damaging hazard globally after storms and earthquakes [20,21].

In this context, North East of Romania is no exception. Crossed by the Siret and Prut rivers, which together constitute the biggest river basins in the country according to their surface, this territory is the most vulnerable area in terms of global climate change and modification of climate patterns. Modifications in the thermal, pluviometric and hydrological regimes have been noticed [22]. In the last 30 years, historical flood events were recorded in this area for the entire territory of Romania [23]. Floods have become a constant threat (especially after the year 2000) as, every two years, a major flood event was documented (e.g., 2004, 2005, 2006, 2008, and 2010) [24–27].

Besides climate change phenomenon implications, the high frequency of floods events is also determined by a whole range of socio-economic factors such as: Land use practices (e.g., changing floodplain functionalities), the development of socio-economic activities in flood-prone areas, changing living standards (e.g., urbanization—construction in high flood risk areas, urban sprawl), and poor regulations reflected by the incapacity of the responsible authorities to implement efficient planning policies [28–32]. The anthropic intervention on land has profoundly modified the natural standard behavior, and the changes on land use and land management are affecting, in particular, the river hydrology that determines the flood hazard [33]. This has ended up being a continuous increased trend in terms of vulnerability and exposure to disasters and flood hazard [34,35]. However, according to [36], the population has, voluntarily and under the pressure of modern society, exposed the environment to the possibility of flooding.

In Romania, there is a lack of proper flood defense measures against floods within many river basins. In general, the current flood defense infrastructure and hydro-technical constructions were built-up in the 1970s and very few improvements have been made in the years after. In the same time, over the past decades, the river regularization measures, the intensification of deforestation in the Carpathian Mountains, and housing development on the floodplains have changed the flood regime in the Siret and Prut river basins and their sub-basins. This has also the case for one of the most modified watercourses in the Siret river basin: Bistriţa River. Ten reservoirs, including canals, dams, collectors, transfer flows, protection works, and banks have been developed on the main course of the Bistriţa River, in order to mitigate the flood effects, and for the production of electricity using hydro-power plants [22,26,27]. Although no significant flood event has occurred on the Bistriţa River due to this considerable protection, the recent changes in pluviometric and hydrological regimes due to climate change, anthropic intervention on land, and reservoir clogging prove to be real challenges in the future flood hazard assessment of this area [22].

According to [37], human societies have always tried to reduce flood impact and have always sought protection from natural disasters by settling in safe areas or by building defense infrastructures. Throughout human history, there has been a constant preoccupation to understand, assess, and predict flood events and their impact [38]. Many ways were introduced in order to deal with flooding phenomena. In this context, aspects like streamflow hydraulic modeling, flood hazard assessment, and flood risk management have received increasing attention in the second half of the 20th century and has become a fundamental issue in the beginning of 21st century. Flood risk management can include: Hazard assessments, exposure assessments, vulnerability assessments, and risk assessments [39,40]. Flood hazard assessment is the most important step in the development of effective flood risk management; the flood hazard maps are used for danger estimation and for taking preventive conservation measures for risk mitigation. In this way, valuable information for different decision makers regarding spatial planning, and the design of infrastructure and emergency response preparation, become available.

In order to conduct flood hazard assessments and propose flood risk management measures, it is important to know what kind of comprehensive tools are available to be used in the analysis of flood susceptibility [41]. Generally, in flood hazard assessment, field-observed data from gauging stations are used. The existence of long records of accurate river discharge measurements is requested. However, these observations can have several uncertainties and the data are not always available [39,42]. Earth Observation (EO) datasets (e.g., space borne, aerial images, and satellite images) together with geographic information systems (GIS) can be used to determine the extent of flood areas and for the production of flood hazard and risk maps [43]. Combined with the use of remote sensing (RS) and GIS, EO datasets provide a favorable environment for relevant information processing in order to obtain the spatial extent of flood hazard areas and flood mapping [44–47]. Even if the benefits are high, this technique offers valuable information only for specific flood events, and this aspect is perceived as a disadvantage because future flood details and impact cannot be investigated [42,45].

Another available tool used in the determination of flood hazard areas is streamflow hydraulic modeling. The usage of this technique witnessed a considerable development in the recent years. As flood events have increased significantly over the last decades under the current unpredictable climate change behavior and human activities, the development of flood inundation models have become a necessity [39]. Streamflow hydraulic modeling is a useful and efficient tool in the simulation of flood events, identification of vulnerable areas, and estimation of spatially-distributed variables such as flow velocity or depth [48]. Since 1970, considerable efforts have been made in order to improve the capabilities and functionalities of this technique. Recent computational advances in hydraulic modeling offer new opportunities to support decision-making and adaptation [49]. The applicability of streamflow models have proved efficient in flood hazard and risk mapping studies [50,51], real time flood forecasting [52], and in remodeling past flood events [53]. Also, the availability of Digital Elevation Models (DEMs) [54] based on high-density Light Detection and Ranging (LiDAR) data have considerably improved the accuracy of flood parameters [47]. Depending on the selected mathematical method and scale, the complexity of models can be classified in simple interpolation methods (1D) and spatially detailed models which solve the water equations in 2D [55]. The use of a 1D simple model is recommended for channels, as long as the water remains in the steep slopes of the channel and no significant water spread is expected. 1D modeling is the perfect solution in cases of flood propagation along the main river [56,57]. Two-dimensional models are recommended in cases when the water is expected to overtop and the flow direction may change, spreading across a large area [42,58]. However, there are many works which make comparisons between 1D and 2D models [56,58–60].

### 1.2. Case Study: Aims and Objectives

In this study, we developed a method for flood vulnerability assessment under real (average discharge) and mathematical (calculated discharge) hydrological data based on HEC-RAS, LiDAR data, and 2D hydraulic modeling. Different from other studies [23–27,47], which provide flood hazard maps based on hydrological data calibrated at river basin scale [23,24,26,27], we developed for the first time in the study area flood hazard maps adapted to local environment settings and calculated discharge. In this context, four 2D streamflow hydraulic scenarios (s1–s4) were computed in order to test the flood mitigation capacity of hydro-technical constructions located downstream on the Bistriţa River (North East Romania) (Figure 1a). The scenarios were based on different discharge releases (average discharge; calculated discharge: s1–s4) from the Bacău 1 reservoir (rB1), also known as the Lilieci storage reservoir, and from its hydro-power plant (H-pp) (Figure 1b). The average and calculated discharges were correlated with the official operating regulations of rB1. The results provided a more realistic perspective about the possible flood threats located downstream: rB1, an area which overlays with the urban and peri-urban area of Bacău City; and rB2, with the Bacău 2 reservoir (Figure 1c).

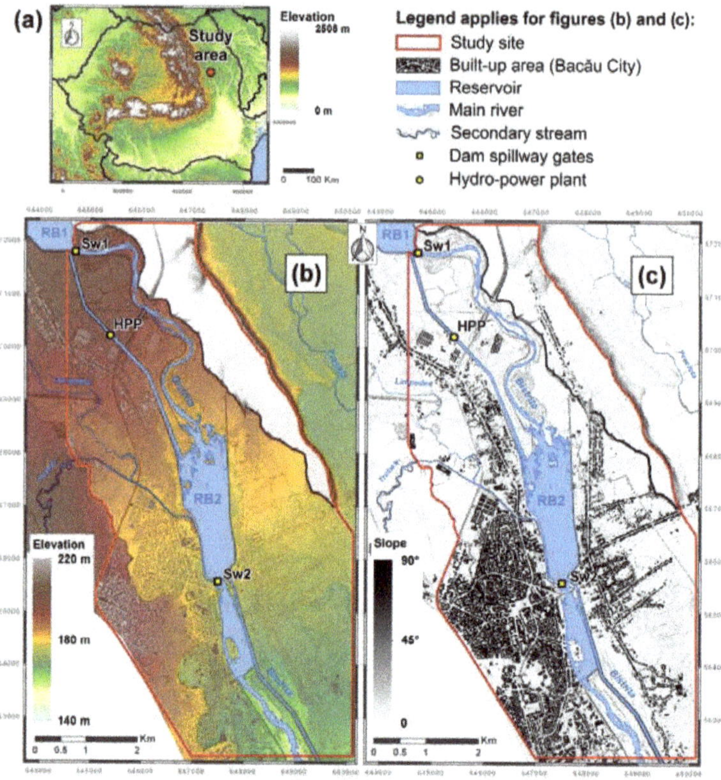


**Figure 1.** (**a**) Geographic location of the study area in North East Romania; (**b**) Elevation map and, (**c**) Digital Elevation Model (DEM)-derived slope map of the urban and peri-urban areas of Bacău City. Data collected by the National Administration Romanian Waters (NARW) in 2014; DEM derived from ground point's elevation data (0.5 point density) collected using Light Detection and Ranging (LiDAR) techniques from aircraft. The abbreviations within the (**b**,**c**) maps show the locations of the study site boundaries used in the 2D hydraulic modeling for the Bistriţa River, Bacău 1 (rB1) and Bacău 2 (rB2) reservoirs, Bacău 1 spillway (Sw1) and Bacău 2 spillway (Sw2) gates, and the Bacău 1 hydro-power plant (H-pp). All spatial data was defined and projected in the Romanian national projection (STEREO 70).

## 2. Methodology

### *2.1. Selected Site for 2D Streamflow Modeling*

The study area selected for 2D streamflow hydraulic modeling and for urban food hazard assessment covers 29.73 km$^2$ in the urban and peri-urban area of Bacău City (Bacău County, North East Romania) (Figure 1a). The town has a population of 196,883 inhabitants according to the 2016 census, making it the 12th largest city in Romania. The considered Bistriţa River reach extends 2 km upstream from the confluence of the Bistriţa and Siret rivers (Figure 1b,c). Due to more than 10 large reservoirs and other complex hydro-technical works (e.g., dams, water gates, hydro-power plants, channels, headraces, flood protection dikes) which equip the downstream sector of the Bistriţa watershed, the predictive flood models are difficult to achieve through classical methods. According to [22], even if the Bacău City was included on the settlements list potentially affected by floods in North East Romania, the flood hazard maps provided by the authorities would not highlight this state. For this reason, in the last 10 years, new neighborhood areas have been built-up in the floodplain of Bistriţa River or in the proximity of the rB1 and rB2 reservoirs.

### 2.2. Data Acquisition

Figure 2 summarizes the workflow chart followed in this study with the hydrological, LiDAR, and built-up data obtaining process, and the key steps for HEC-RAS 2D streamflow and flood modeling.

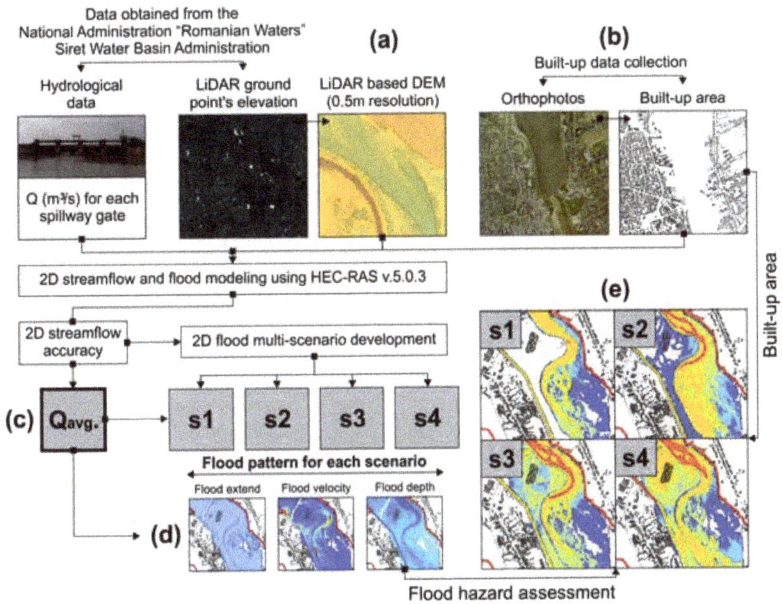


**Figure 2.** Workflow chart of 2D streamflow and flood hydraulic modeling process using hydrological data (Sw1) and LiDAR ground point elevation (0.5–3 m spatial density) within the urban and peri-urban area of Bacău City: (**a**) Generate the LiDAR based DEM with 0.5 m spatial resolution based on LiDAR ground point elevation; (**b**) manual digitalization of buildings using a high resolution orthophotos and built-up data integration in DEM; (**c**) generate the four 2D multi-scenario (s1–s4) using HEC-RAS v.5.0.3, where $Q_{avg.}$ was generated based on average discharge at Sw1 for 2D streamflow accuracy, and s1–s4 was computed using calculated discharge at Sw1 for 2D flood multi-scenario development; (**d**) generate the flood pattern for each computed scenario and export the individual layers (e.g., flood extent, flood velocity, flood depth); (**e**) flood hazard assessment based on the built-up data and flood depth classification according to the Ministry of Land Infrastructure and Transport (MLIT) [42].

#### 2.2.1. Development of DEM

The DEM data used in this study comprehends airborne LiDAR technology. LiDAR DEMs display the terrain with a high degree of horizontal and vertical accuracy, a feature that is essential to terrain-related applications such as streamflow and flood modeling [54,61,62]. The data was obtained from the National Administration Romanian Waters (NARW)–Siret Water Basin Administration (SWBA) and consisted of raw ground point elevation data, at a spatial density between 0.5 and 3 m. More than $7.833 \times 10^6$ points covering almost 100 $km^2$ was considered in the DEM computation (Figure 2a). The final product (DEM) was obtained by means of *natural neighbor* interpolation technique [63]. The *natural neighbor* interpolation technique was preferred as it finds the closest subset of input samples to a query point and applies weights to them based on proportionate areas in order to interpolate a value [64]. To capture as many topographical details as possible, the DEM was computed at 0.5 m spatial resolution (Figure 2a).

As the considered study area overlaps a highly urbanized space, crossed by one of the most developed hydro-technical rivers in the country, an overview regarding the current situation of the urban built-up areas was necessary. Initially the open source database offered by OpenStreetMap [65] was consulted. As the available database proved incomplete, the last step in built-up data acquisition consisted of the manual digitalization of the missing built-up areas using a high resolution orthophotos digital image (0.5 m resolution), dating from 2012. In this way, the minimum and maximum polygonal areas of the improved built-up vector was situated between 10 $m^2$ and 19,534 $m^2$ (Figure 2b). Furthermore, in order to assure the hydraulic streamflow accuracy, the obtained vector was integrated into the DEM. The new data was rasterized and then added to the already obtained LiDAR DEM, taking into consideration an average height of 10 m. The built-up areas affected by the inundation were identified using the built-up vector obtained in the previous steps. All residential and industrial built-up areas that intersected or shared a common spatial extend with the inundation extend, were considered impacted (Figure 2b).

#### 2.2.2. Hydrological Data

The hydrological data used in this study was extracted from the Lilieci reservoir (rB1 and H-pp) official operating rules dated from 2012 and offered by SWBA (Table 1). According to Romanian regulations STAS—4273/61 and STAS 4273/82, the Lilieci reservoir (rB1) is classified after the height of the dam and its storage capacity volume, in the third importance category—having the flow sizing rate Q2% (50-years recurrence interval) = 510 $m^3/s$, and Q0.5% (500-years recurrence interval) verification flow rate of 1670 $m^3/s$. Nevertheless, the rB1 spillway gates (Sw1) and levees were designed to transit flow rates that correspond the second importance category—Q1% (100-years recurrence interval) = 1140 $m^3/s$ and Q0.1% (1000-years recurrence interval) = 2100 $m^3/s$. The H-pp is located at approximately 1726 m from the Sw1, on a specially designed channel and it is equipped with two Kaplan turbines with a 180 $m^3/s$ maximum flow rate.

**Table 1.** The volume of water contained (million $m^3$) in the rB1 reservoir according to specific water surface elevations (m) and related scenarios (s1–s4) developed in the study.

| **Water surface elevation (m)** | 172.14 | 177.86 | 178.14 | 178.64 | 179.14 | 179.21 | 179.64 | 179.8 | 180.14 |
|---|---|---|---|---|---|---|---|---|---|
| **Volume of water contained (million $m^3$)** | 0 | 4.8 | 5.41 | 6.77 | 8.18 | 8.4 | 9.66 | 10.1 | 11.18 |
| **Related scenario** | | s1 | | s2 | | s3 | | s4 | |

### 2.3. *2D Hydraulic Modelling*

In relation to watercourses (e.g., streams, rivers, channels), the hydraulic models describe water movement through space in three directions [66]: A downstream direction along the river channel, a lateral direction (e.g., whenever the water begins to spill out overland), and a vertical direction which practically defines the height of the flood. Standard flood modeling practices include 1D modeling (upstream to downstream direction), 2D modeling (downstream and lateral directions), hybrid 1D–2D modeling and 3D numerical models, along with hydrograph design, specified ground roughness, and accurate digital elevation data [39,67]. For the present study, a 2D approach was adopted for a number of reasons (Figure 2c):

- The main purpose of the study was to develop overtopping multi-scenarios in order to define the areas with risk of flooding in a complex urban environment; where it cannot be assumed that all flow will be parallel to the main river and where a higher hydrodynamic accuracy is required [68] (Figure 2d,e).
- Considering that the water will not propagate in the vertical direction, the vertical water velocities will be negligible compared to the horizontal velocities and also the kinetic energy losses due to the vertical movement. That is why this study has considered a 2D model to develop its conclusions.
- The highly accurate representation of the river's bathymetry and floodplain topography, which is represented by a high resolution digital elevation's model.

- Two-dimensional flood propagation modeling offers additional information regarding some characteristics of the flood, such as flow velocity and water trend propagation.

The 2D functionality builds flood models in a more accurate way, considering the flow variability in time, and both spatial dimensions, the $x$ and $y$, making it more suitable for case studies that consider wide floodplains, urban environments, dam/levee breach situations, where the water is expected to spread over an open, unconstrained area, in multiple directions [68]. The release of the HEC-RAS software (5.0 version), in 2016, integrated 2D unsteady flow capabilities, offering the possibility to analyse water propagation over a predefined surface, which is found in the form of a digital elevations model [69,70]. HEC-RAS is a well-known software, capable of modeling a flood inundation event [39,42]. The scenarios proposed by this study were simulated using the open source HEC-RAS software (5.0.3 version), developed by U.S. Army Corps of Engineers (USAGE). Starting with HEC-RAS software version 5.0, two-dimensional unsteady water flow modeling can be performed. The program 2D flow modeling algorithm solves either shallow water equations, also called bidimensional Saint Venant equations (Equation (1)), or the 2D diffusion wave equations (Equations (2) and (3)).

$$\frac{\partial \zeta}{\partial t} + \frac{\partial p}{\partial x} + \frac{\partial q}{\partial y} = 0 \quad (1)$$

$$\frac{\partial p}{\partial t} + \frac{\partial}{\partial x}\left(\frac{p^2}{h}\right) + \frac{\partial}{\partial y}\left(\frac{pq}{h}\right) = -\frac{n^2 pg\sqrt{p^2+q^2}}{h^2} - gh\frac{\partial \zeta}{\partial x} + pf + \frac{\partial}{\rho \partial x}(h\tau_{xx}) + \frac{\partial}{\rho \partial y}\left(h\tau_{xy}\right) \quad (2)$$

$$\frac{\partial q}{\partial t} + \frac{\partial}{\partial y}\left(\frac{q^2}{h}\right) + \frac{\partial}{\partial x}\left(\frac{pq}{h}\right) = -\frac{n^2 qg\sqrt{p^2+q^2}}{h^2} - gh\frac{\partial \zeta}{\partial y} + qf + \frac{\partial}{\rho \partial y}\left(h\tau_{yy}\right) + \frac{\partial}{\rho \partial x}\left(h\tau_{xy}\right) \quad (3)$$

where, $h$ is the water depth (m), $p$ and $q$ are the specific flow in the $x$ and $y$ directions ($m^2s^{-1}$), $\zeta$ is the surface elevation (m), $g$ is the acceleration due to gravity ($ms^{-2}$), $n$ is the Manning's Roughness coefficient ($m^{-1/3}$ s), $\rho$ is the water density (kg $m^{-3}$), $\tau_{xx}$, $\tau_{yy}$, and $\tau_{xy}$ are the components of the effective shear stress and $f$ is the Coriolis ($s^{-1}$). When the diffusive wave is selected, the inertial terms of the momentum equations are neglected: $\partial p/\partial t + \partial/\partial x\left(p^2/h\right) + \partial/\partial y\ (pq/h) = 0$ (Equation (2)); $\partial q/\partial t + \partial/\partial y\left(q^2/h\right) + \partial/\partial x\ (pq/h) = 0$ (Equation (3)) [42,45,71].

Due to their faster computational time and greater stability properties, the 2D diffusion wave equations were preferred in the present study [70]. In this case, the floodplain flow was approximated as a two-dimensional diffusion wave where water can flow in any direction based on the defined topography and the resistance to flow determined by the type of land use [72,73]. A 2D model consists of a 2D computational grid which groups/discretizes the river and its adjacent areas into a collection of individual and connected cells, called grid cells (2D flow cells), that are used to characterize the underlying topography [56,74]. The 2D flow area is the region (the boundary selected by the user) where the water flow will be modeled.

A close polygonal mesh with a computation point spacing of 5 m was built in order to define the spacing between the computational grid-cells. This managed to capture all the LiDAR terrain characteristics. In this way, more than 2,000,000 grid cells were generated; therefore, no break line was needed to be introduced (e.g., top of the levees, along the main channels). The computational mesh controlled the water flow throughout the 2D flow designed area and from cell to cell. Considering the underlying terrain and the generated computational mesh, the software developed detailed elevation–volume relationships and detailed hydraulic property curves for each cell face (elevation vs. wetted perimeter, area, and roughness) [70]. In this way, the water elevation values were calculated for each centroid of the grid. The Manning's Roughness coefficients, specific to each landcover class was set according to [75] (Figure 3).

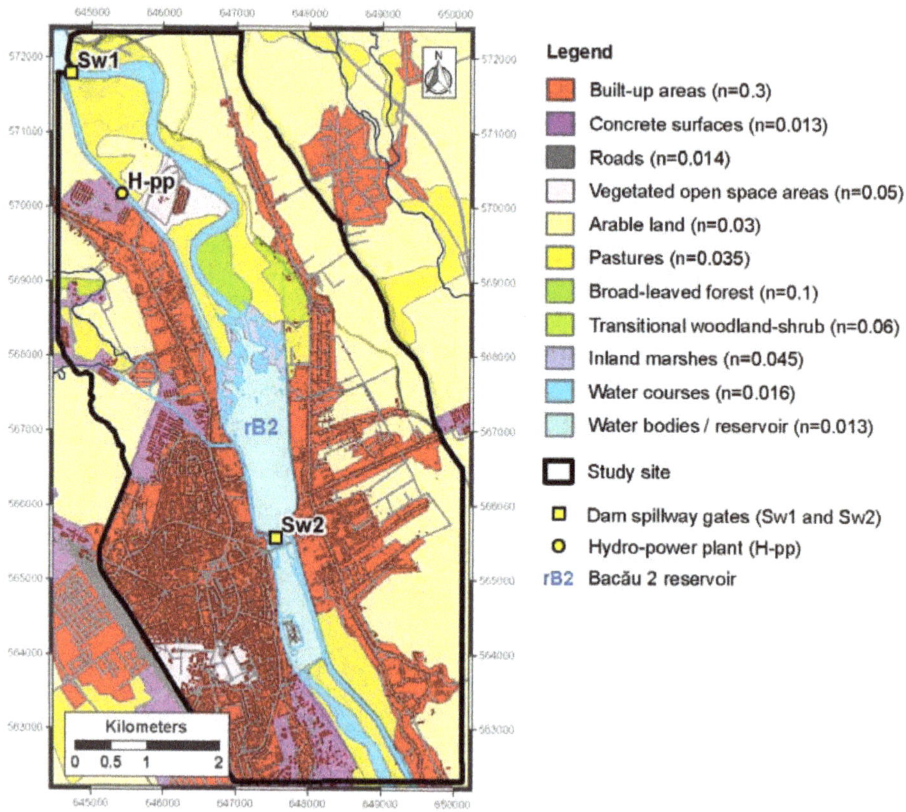


**Figure 3.** Landcover map within the study area and Manning's Roughness coefficient $n$ ($m^{-1/3}$ s) value for each class according to [75].

After the geometric data is created and the roughness coefficient is set, the next step consists of establishing the boundary conditions. This means that flood data along the boundaries of the 2D flow area is assigned. Flow hydrographs and normal depth boundary conditions representing the average riverbed slope are used in order to bring water into the 2D flow area. Four flood scenarios were simulated and, for each scenario, two hydrographs and two normal depths boundary conditions were introduced. The hydrographs were 24 h long with values recorded each hour. The upstream boundary conditions (the hydrographs representing imposed flow condition) were located in two specific locations downstream of Lilieci dam: At the hydropower plant of Lilieci reservoir (on the controlled channel), and on the natural river bed (where water from the reservoir is released). The imposed water level boundary condition was located at the downstream extremes of the channel and at the end of storage lake control channel (in the limits of DEM). An energy slope of $10^{-4}$ $mm^{-1}$ was used.

In order to ensure the stability of the model, the time steps were estimated according to the Courant–Friedrichs–Lewy condition [70] (Equation (4)).

$$C = \frac{V\Delta T}{\Delta x} \leq 1.0 \text{ (with } C_{max} = 3.0) \text{ or } \Delta T \leq \frac{\Delta x}{V} \text{ (with } C = 1.0) \quad (4)$$

where, $C$ is the Courant number, $V$ is the flood wave velocity (m/s), $\Delta T$ the computational time step (s), and $\Delta x$ the average cell size (m) [67]. According to Equation (4), a time step of 10 s was selected. To run a 24 h simulation at a time step of 10 s, the 2D models took between ~3 h (for $Q_{avg}$. scenario generated based on average discharge) and ~12 h (for each s1–s4). The flood inundation (depth), flood velocity, and other related results were obtained for each hour according to the hydrograph output interval.

*2.4. Streamflow and Flood Simulation Accuracy*

For accuracy purposes, a 2D water flow test simulation was computed. The data needed for this flow accuracy assessment test consisted of average discharge values at the Lilieci reservoir (rB1) for the year 2012 and, respectively, an orthophoto digital image (Figure 4a) and a Landsat 5 satellite image (Figure 4b) showing the water extend in the area acquired in the same date. The results of this simulation showed a close agreement between the water extend obtained from the model computation and the available orthophoto digital image (Figure 4c). The performance assessment was realized by comparing these three sets of spatial data. A difference of 5% (16 ha) was recorded between the orthophoto digital image water extend and the 2D average discharge simulation result. A second comparison was carried out using the Landsat 5 satellite image, acquired on 26 August 2012. The water extend were extracted according to [23] and compared with the 2D average discharge simulation result. In this case a difference of 9% (29 ha) was observed. Taking into consideration the presence of vegetation which could have masked the presence of water in the digital orthophoto images and the coarse resolution of Landsat scene, the accuracy of the flood extent assessment was judged satisfactory.

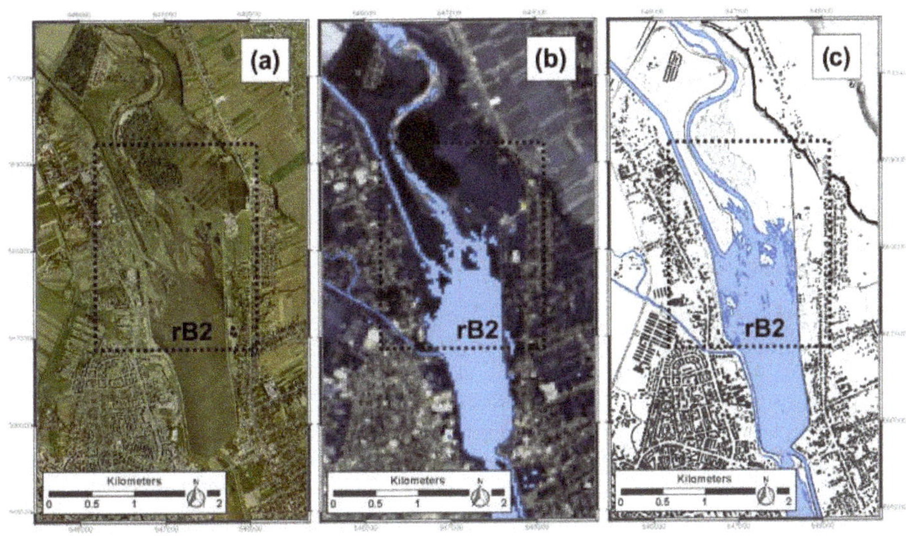


**Figure 4.** Streamflow 2D accuracy: (**a**) orthophoto (digital image) for 2012, (**b**) Landsat 5 (satellite image) for 2012, and (**c**) results of 2D average discharge simulation result. In all images was highlighted the water extend within inflow area (fan-delta) of the rB2 reservoir for comparison.

*2.5. Multi-Scenario Development*

The flood scenarios were based on hydrological data extracted from the Lilieci reservoir (rB1) official operating rules. Upstream boundary conditions for all scenarios were developed on a simulation of water flow over a period of 24 h, at a temporal resolution of 1 h. The average discharge scenario, spatially displays the main channel at a water release of Q = 34 $m^3$/s located at the Lilieci hydro-power plant (H-pp) and a Q = 2.8 $m^3$/s flow located at the reservoir main gates (Sw1), where the water is released directly in the main river bed (in order to assure the river servitude discharge). Scenario 1 (s1) takes into consideration the average discharge at H-pp of the Lilieci reservoir (Q = 34 $m^3$/s) and the maximum water release in the case of the full opening of one gate (Sw1) of the same storage lake (Q = 490 $m^3$/s). The maximum values were used in the simulation for a period of 1 h, whereas for the remaining 23 h, the flow rates were set to their average values (Q = 2.8 $m^3$/s). Scenario 2 (s2) considers water flow modeling of Q = 34 $m^3$/s at the H-pp and Q = 980 $m^3$/s at the Sw1 when two gates are fully opened. The distribution of time periods was the same as for s1. For scenario 3 (s3), the same parameters were used as for s1 and s2, with the exception that a water flow of Q = 1470 $m^3$/s is assumed when all 3 gates of the Sw1 are opened. The last scenario 4 (s4) simulated the water flow when all 4 gates of the Sw1 are fully open at a discharge rate of Q = 1960 $m^3$/s. The other parameters remained unchanged. In Table 2 are the simulation parameters for the 4 modeled scenarios.

**Table 2.** Streamflow features and flood multi-scenario developed in this study.

| 2D Streamflow Scenario Code | No. of Open Gates (Sw1 [1] and H-pp [2]) | No. Hours/Open Gates (Sw1 [1]) | No. Hours/2D Simulation | Q [3] ($m^3$/s) (Sw1 [1] + H-pp [2]) | Total Q [3] ($m^3$/s) | Q [4] (%) from Sw1 [1] |
|---|---|---|---|---|---|---|
| Avg. discharge | Servitude discharge and H-$pp_{avg}$ | 24 h | 24 h | 2.8 + 34 | 36.8 | - |
| Scenario 1.–s1 | 1 and H-$pp_{max}$ | 1 h | 24 h | 490 + 34 | 524 | 25 |
| Scenario 2.–s2 | 2 and H-$pp_{max}$ | 1 h | 24 h | 980 + 34 | 1014 | 50 |
| Scenario 3.–s3 | 3 and H-$pp_{max}$ | 1 h | 24 h | 1470 + 34 | 1504 | 75 |
| Scenario 4.–s4 | 4 and H-$pp_{max}$ | 1 h | 24 h | 1960 + 34 | 1994 | 100 |

[1] Sw1: Bacău 1 spillway gates; [2] H-pp: Hydro-power plant; [3] Q: Discharge ($m^3$/s); [4] Q: Percentage of total spillway gates discharge capacity.

## 3. Results

### 3.1. Flood Pattern

The 2D modeling was performed using the HEC-RAS 5.0.3 version which offered various output possibilities in terms of detailed animation and mapping of flood characteristics within the RAS mapper feature. Following the calculation of the 4 scenarios given in Table 1, individual layers of the flooded areas along Bistriţa River and Bacău City regarding the flood extent, velocity, water surface elevation and depth, were exported.

#### 3.1.1. Flood Extent

The flood extent layer consists of an *Inundation Boundary* shapefile vector layer capturing the areas affected by the flood during the whole 24 h simulation period (Figure 5). According to s1, an area of 6.27 $km^2$ within the Bistriţa floodplain, the inflow area (fan-delta) of rB2 and downstream sector of rB2 reservoir, are potentially affected by floods. Also, six buildings located within the s1 flood extent are potentially affected. In the s2 simulation, the total flood extent occupies 8.66 $km^2$ and 173 buildings in the peri-urban area of Bacău City which are located in the flood zone. In the third scenario (s3), the water release from Sw1 and H-pp spills over protection dikes and floods the built-up area of Bacău City. The total flood extent increases to 11.46 $km^2$ (built-up area—1.09 $km^2$) and 2461 of buildings (e.g., block with flats, houses and yards, other residential buildings, industrial units) are potentially affected by the inundation waters. The last scenario (s4) indicates the most catastrophic situation where total flood extent is 12.82 $km^2$, from which 1.5 $km^2$ are built-up area, and 3780 residential buildings are vulnerable to the flood hazard (Table 3). Overall, in the first two scenarios (s1 and s2), the flood extent affected only the built-up areas within the Bistriţa floodplain, and in the last two scenarios (s3 and s4) the flood extent occupied the peri-urban and urban areas.

**Table 3.** Flood extent area, built-up area, and number of buildings potentially affected by floods computed for each scenario (s1–s4).

| Flood Extent | s1 | s2 | s3 | s4 |
|---|---|---|---|---|
| Total flood extent ($km^2$) | 6.27 | 8.66 | 11.46 | 12.82 |
| Built-up area ($km^2$) | 0.001 | 0.04 | 1.09 | 1.5 |
| Number of affected buildings | 6 | 173 | 2461 | 3780 |

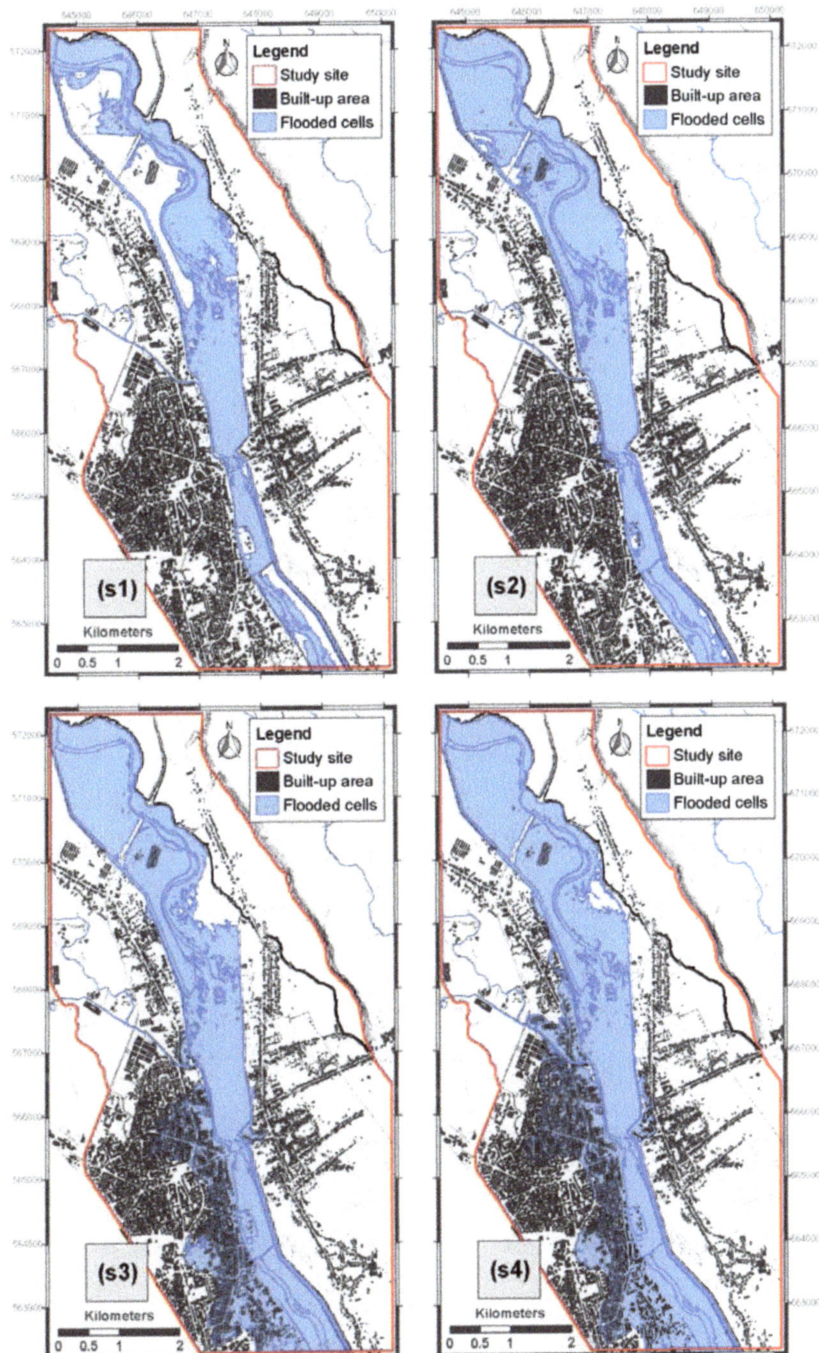


**Figure 5.** Flood extent derived from the HEC-RAS 2D multi-scenario generated based on LiDAR data and average discharge; s1–s4 was computed using calculated discharge (see Table 2).

### 3.1.2. Flood Velocity

Another important result of the 2D modeling process is flood velocity. This is computed by recording the maximum velocity for each cell in the computational mesh regardless the time when it was measured in the whole 24 h simulation period (Figure 6). According to all computed scenarios (s1–s4), the 0.01–1 m/s velocity class has the highest frequency (88.3%), which corresponds with the average velocity recorded downstream rB2 (0.6 m/s). In the first two scenarios (s1 and s2), all vulnerable buildings are potentially affected by the 0.01–1 m/s flood velocity class, except three buildings located in front of Sw1 or along the natural watercourse of Bistriţa River. In the case of s3, only 9% of total affected buildings are threatened by 2 m/s flood velocity, and in case of s4, only 14% of total built-up area is potentially affected by high velocity of water (1–5 m/s). However, due to floodplain roughness and hydro-technical works within study area, the high flood velocities (>5 m/s) are registered only at the Sw1 and Sw2 gates, and through the narrow sections between rB1 and rB2 river sector (Table 4). Overall, in each computed scenario, the flood velocity is not the main threat for built-up area, but the river morphology (e.g., river bed, banks, alluvial deposits), and the hydro-technical works conservation stage (e.g., dams, channel, bridges) can alter through time. For these reasons, the maintenance of the flood mitigation equipment within urban and peri-urban areas must be a priority for the competent authorities.

**Table 4.** Number of buildings potentially affected by floods vs. flood velocity (m/s) computed for each scenario (s1–s4).

| Flood Velocity (m/s) | s1 | s2 | s3 | s4 |
|---|---|---|---|---|
| <1 | 6 | 170 | 2260 | 3238 |
| 1–2 | | 3 | 184 | 353 |
| 2–3 | | | 17 | 189 |
| 3–4 | | | | |
| 4–5 | | | | |

### 3.1.3. Flood Depth

The maximum flood depth maps were generated by the 2D model by taking into consideration the maximum depth for each cell, no matter the time when that maximum depth was registered during the whole 24 h simulation period (Figure 7). According to s1 and s2, 95.5% of vulnerable buildings are potentially affected by a flood depth which does not exceed 1 m depth. In case of s3, when water levels exceed the height of protection dikes and inundates the urban and peri-urban area of Bacău City, 1174 buildings will be affected by floods which do not exceed 1 m in depth; 39.4% of the vulnerable built-up areas are potentially affected by floods with a depth between 1–2 m, and 12.7% by floods with a depth between 2–3 m. In the same scenario (s3), only 4 buildings are located in areas with water that exceeds 3 m. In case of s4, the most catastrophic scenario taken into account in this study, 34.7% of buildings would be affected by floods with <1 m depth, 30.9% by floods with depths between 1–2 m, 25.9% by floods with depths between 2–3 m, 7.9% by floods with depths between 3–4 m, and the rest of 0.4% (18 buildings) by floods that exceed >5 m depth (Table 5).

**Table 5.** Number of buildings potentially affected by floods vs. flood depth (m) computed for each scenario (s1–s4).

| Flood Depth (m) | s1 | s2 | s3 | s4 |
|---|---|---|---|---|
| <1.0 | 6 | 165 | 1174 | 1312 |
| 1.0–2.0 | | 8 | 971 | 1171 |
| 2.0–3.0 | | | 312 | 980 |
| 3.0–4.0 | | | 4 | 299 |
| 4.0–5.0 | | | | 18 |

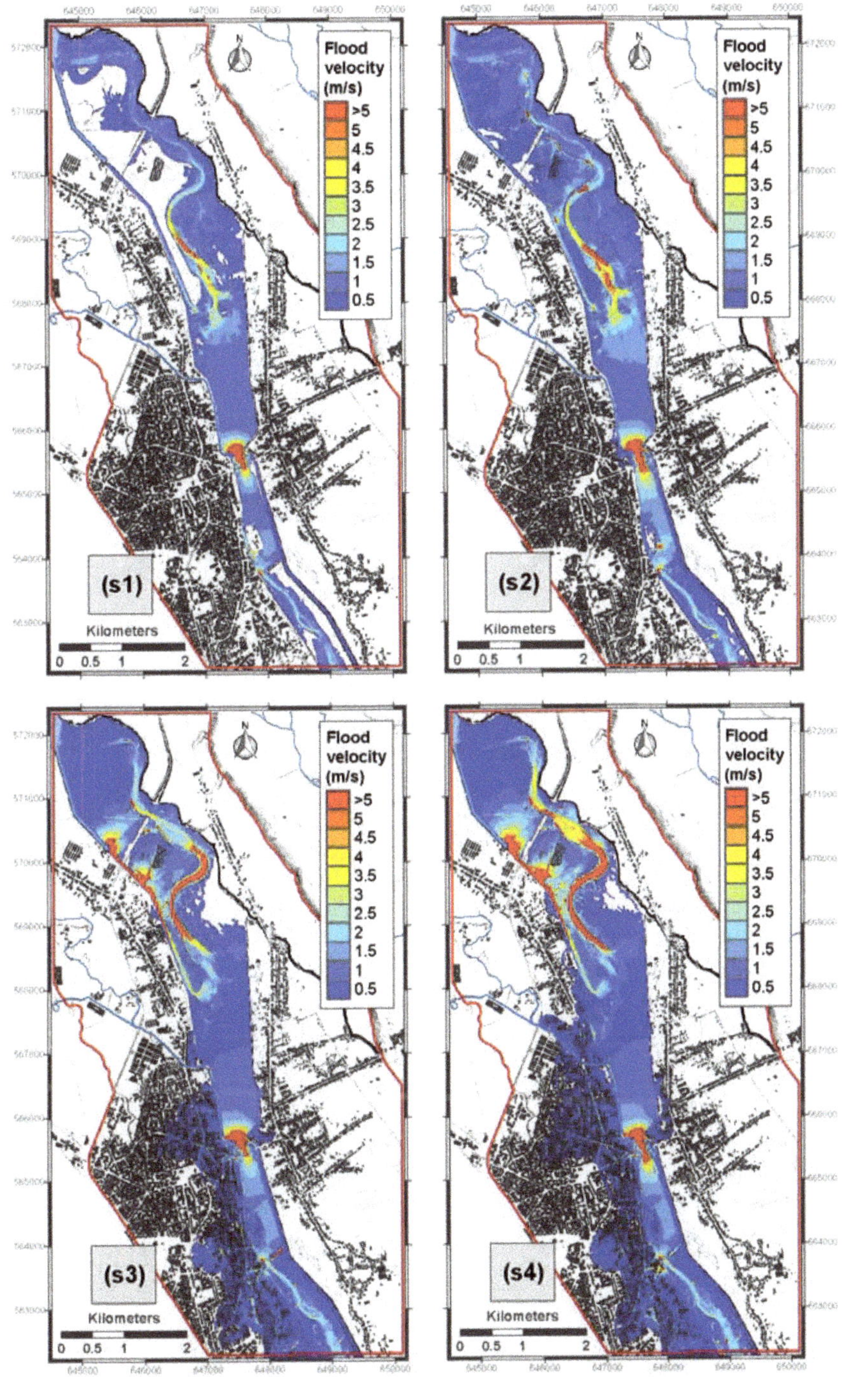


**Figure 6.** Flood velocity derived from the HEC-RAS 2D multi-scenario generated based on LiDAR data and average discharge; s1–s4 was computed using calculated discharge (see Table 2).

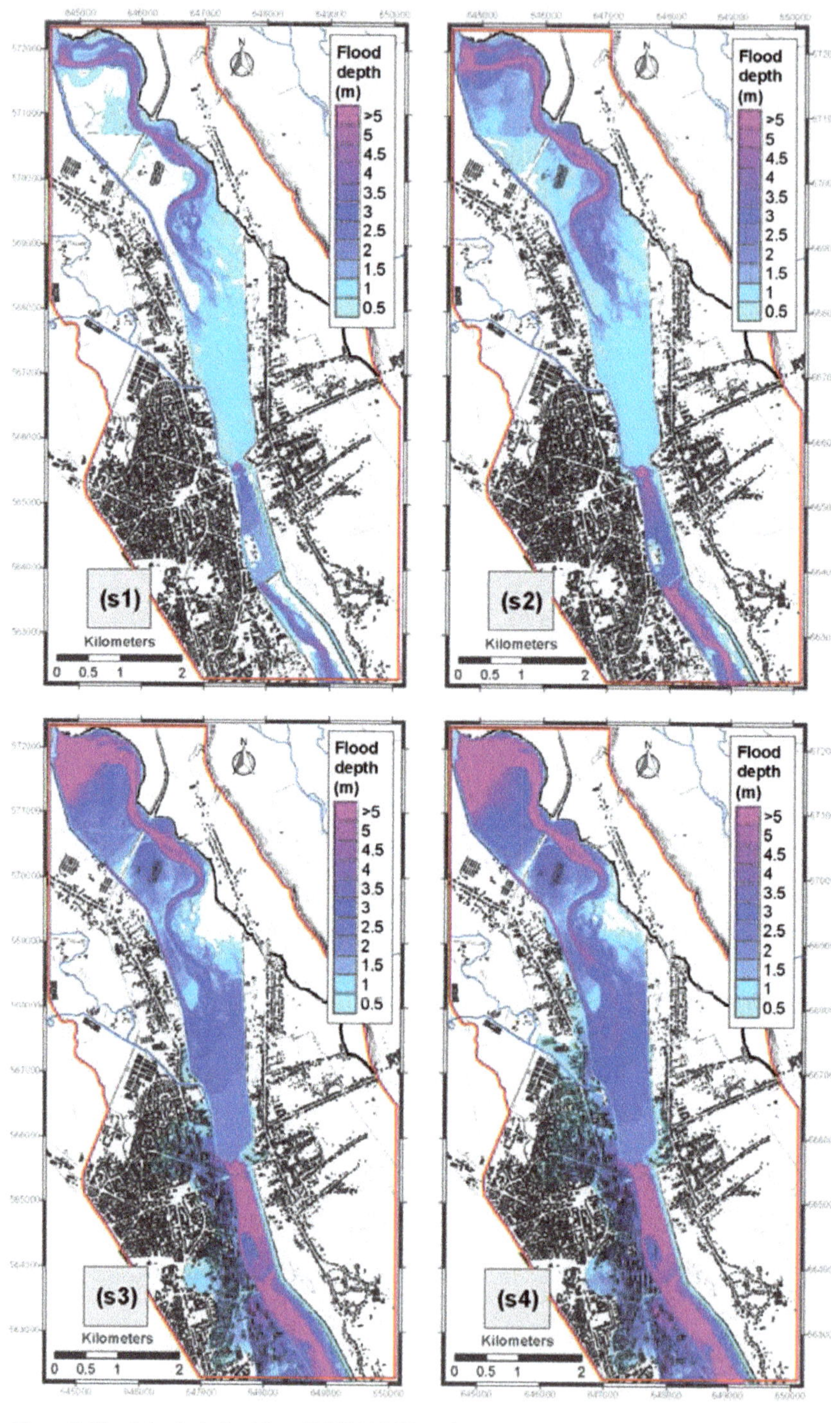


**Figure 7.** Flood depth derived from HEC-RAS 2D multi-scenario generated based on LiDAR data and average discharge; s1–s4 was computed using calculated discharge (see Table 2).

### 3.2. Flood Hazard Assessment

Usually, the flood hazard assessment is based on quantifiable variables like flood extent, water velocity or water depth, and indicate the vulnerability of built-up areas to hydrological events with possible destructive impact. In this study, we provided the flood hazard assessment using only the flood extent and flood depth resulting from the four HEC-RAS 2D scenarios (s1–s4). The flood velocity was not taken into account because is a constant variable in all four flood scenarios (88.3% from total built-up area are affected by the 0.01–1 m/s velocity class). To generate the flood hazard categories, the water depth for each flood extent was classified according to the Japanese criteria of the Ministry of Land Infrastructure and Transport (MLIT) [42]. The criteria suggest five flood hazard categories: H1—very low hazard (flood depth < 0.5 m); H2—low hazard (flood depth between 0.5–1 m); H3—medium hazard (flood depth between 1–2 m); H4—high hazard (flood depth between 2–5 m); H5—extreme hazard (flood depth > 5 m) (Figure 8). The flood hazard classification methodology and hazard description based on flood depth is detailed in Table 6.

**Table 6.** Flood hazard classification based on water depth according to the MLIT [42].

| Flood Hazard | Flood Depth (m) | Hazard Classes | Hazard Description |
|---|---|---|---|
| H1 | <0.5 | Very low | Flood does not pose hazard to people and on-foot evacuation is not difficult. |
| H2 | 0.5–1 | Low | Flood water poses hazard for infants and on-foot evacuation of adults becomes difficult; evacuation becomes more complicated. |
| H3 | 1–2 | Medium | Flood depth can drown people; people may be safe inside their homes. |
| H4 | 2–5 | High | People are exposed to flood hazard even inside their homes and evacuate towards the roof of their homes is suggested. |
| H5 | >5 | Extreme | Built-up structures like homes may get covered by the flood; people may get drowned even if they evacuate towards the roof of their homes. |

According to s1, even if all 5 hazard classes are encountered, all vulnerable buildings are situated in the very low class of hazard (H1). In case of s2, 86.7% of total vulnerable buildings are very low affected by flood (H1), 8.75% are low affected by flood (H2), and the rest of 4.6% are located in the medium exposed area to flood hazards (H3). In s3 situation, the number of potentially affected buildings is significantly increased: H1—674 buildings; H2—500 buildings; H3—971 buildings; and H4—316 buildings. Also, this is the moment when the floods will affect the urban area of Bacău City. According to s4, more the 34.3% of total flood extent are located in the high (H4) and extreme (H5) hazard classes, the number of potentially affected buildings exceed the all NARW scenarios based on recurrence flood interval probabilities: H1—694 buildings; H2—618 buildings; H3—1171 buildings; H4—1279 buildings; and H5—18 buildings (Table 7; Figure 9a,b).

**Table 7.** Number of buildings potentially affected by floods vs. flood hazard classes computed for each scenario (s1–s4).

| Flood Hazard Classes [1] | s1 | s2 | s3 | s4 |
|---|---|---|---|---|
| H1 (Very low) | 6 | 150 | 674 | 694 |
| H2 (Low) | | 15 | 500 | 618 |
| H3 (Medium) | | 8 | 971 | 1171 |
| H4 (High) | | | 316 | 1279 |
| H5 (Extreme) | | | | 18 |

[1] Flood hazard classes according to [42] (see Table 6).

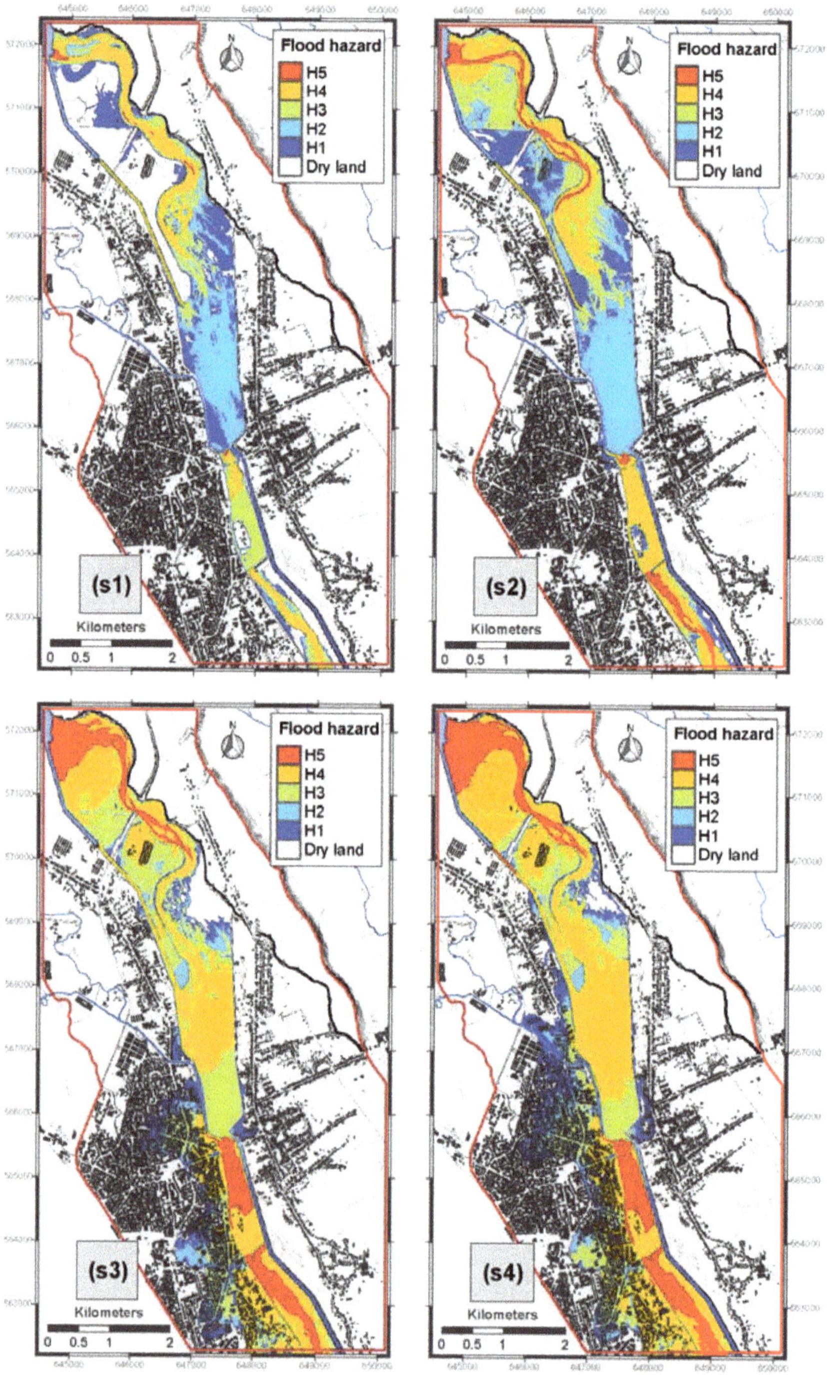


**Figure 8.** Flood hazard maps based on flood depth classification according to the MLIT [42] (see Table 5). Data derived from HEC-RAS 2D multi-scenario generated based on LiDAR data and average discharge; s1–s4 was computed using calculated discharge (see Table 2).

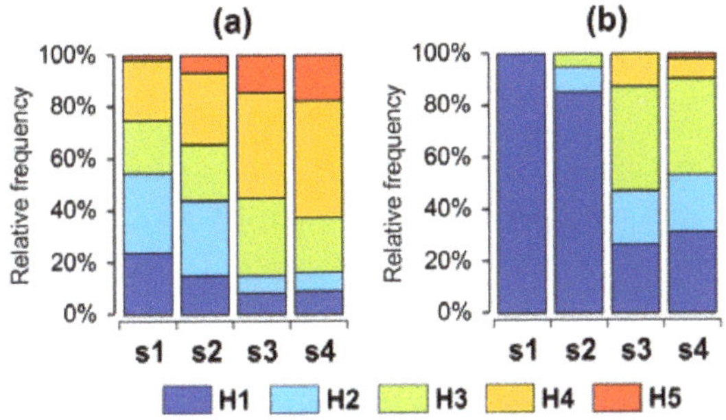


**Figure 9.** The distributions of flood hazard classes within (**a**) flood extent area, and (**b**) built-up area, according to MLIT [42] criteria (see Table 6). The statistics apply for all 2D multi-scenarios (s1–s4) in each chart.

## 4. Discussion

Floods can cause a tremendous amount of damage, especially in the highly regulated river sectors overlaid with highly urbanized areas along the floodplain. For this reason, the multiple viewpoints, methods, assumptions, and future possibilities adapted to new trends (social, economic, natural) and determined by different factors, play an important role in the flood hazard management and establishing flood-vulnerable areas [39]. Even in a highly regulated river system, flood hazards exist; properties and lives are in danger. A flash flood, dense precipitation, or an error in the discharge flow at the gates of the reservoirs can turn into a catastrophic flood event. For this reason, a good preparation regarding this topic is always welcomed. As remote sensing (RS) techniques continue to improve and the availability of data increases, more RS data will be integrated and used in flood modeling. In this context, according to [54], 2D LiDAR DEM based flood simulations provide the best results. The horizontal, vertical accuracy is better on flat terrain (such as wide floodplains) and the integration of vegetation and building heights offer the perfect support for flood hydraulic simulations and accurate water flow propagation.

Due to the unavailability of LiDAR surface elevation points, we were not able to perform the simulation on a Digital Surface Model (DSM with vegetation and building integration). We computed the DEM and manually integrated the building heights. Manually integration of building heights (in the case of data unavailability) can be a solution. Prior to this step, we wanted to check if any open source satellite imagery could be used in order to extract the built-up areas. Initially, a Landsat 8 satellite image dated 11 August 2018 was downloaded using the US Geological Survey (USGS) website [76], pre-processed and processed—including radiometric and atmospheric calibration, maximum likelihood classification (MLC), classification accuracy assessment—in order to extract the areas containing artificial surfaces, residential, commercial, industrial, and transportation infrastructure. A full methodology of the previous mentioned steps can be consulted in the works of [77]. In this way a full perspective regarding the current situation of the urban development throughout the floodplain can be obtained. However, even though this method is practical and can be used in areas where no built-up data is available, the results proved to be at a coarse resolution and impracticable for the level of study we proposed. Therefore we decided to extract the built-up areas at a higher resolution. We manually digitized the missing built-up areas using an orthophotos digital image, dating from 2012. The association of open source imagery (Landsat, Sentinel) and LiDAR data offered unsuitable results.

The 2D flow models and LiDAR DEMs prove to be the perfect combination when it comes to the flood hazard assessment and accurate inundation delineation. Due to its accurate representation of the complex hydraulic conditions that can be found in floodplains (e.g., channels, confluences, bridges, water reservoirs, roads, etc.), 2D hydraulic models can capture the hydraulic behavior of the river in a more accurate way. Water propagation, extent, velocity, and elevation can be reproduced under real and hypothetical flow data and boundary conditions input. In this way we can simulate for different purposes the past flood events, using registered hydrographs at gauging stations, or we can imagine a theoretical hypothesis and, in this way, improve flood hazard and risk maps. This method is the best option in case of flood map creation due to possible reservoir dam failure at various dam discharges. On the other hand, the unavailability of high resolution DEMs can end up underestimating the flood hazard and its dangers. A precise representation of reality is mandatory for accurate flood mapping process.

Our study was focused on a highly regulated and well-equipped hydro-technical river sector which overlays the urban area of Bacău. This proved to be an impediment in the validation process of the 2D hydraulic modeling, as no notable flood event was recorded here in the recent years. Controlled water discharges from the reservoirs located upstream protect the areas downstream. For this reason, a considerable development of built-up areas and urban sprawl was noticed downstream, on both sides of the riverbed. Having only the average official discharges for the year 2012 available at the H-pp and Sw1 to model, we still managed to model the flood hazard in the area based on the official operating rules of the dam Sw1 and taking into consideration the reservoir maximum water retention capacity (Figure 10a).

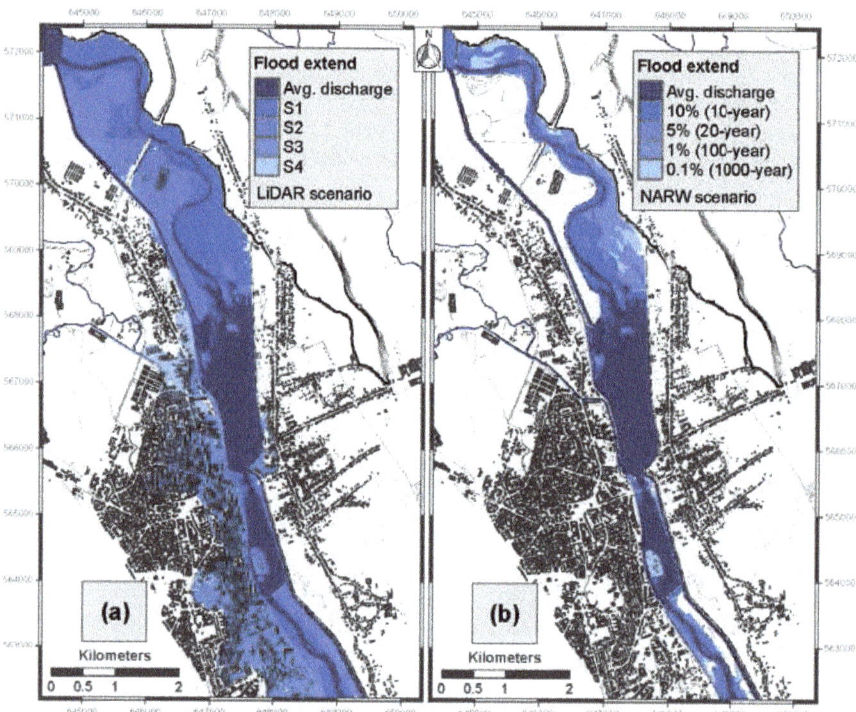


**Figure 10.** Comparison between flood extent maps within the urban and peri-urban area of Bacău City: (**a**) 2D HEC-RAS multi-scenario based on average discharge and calculated discharge (s1–s4) developed in these study, and (**b**) official flood extent maps (NARW) [78] based on large-scale hydrological and spatial data, calculated for 10% (10-year), 5% (20-year), 1% (100-year), and 0.1% (1000-year) recurrence intervals.

Furthermore, we managed to compare the 2D flood extent results with the official available hazard and risk maps available in [78], which was realized under the Directive 2007/60/EC of the European Parliament and Council [79,80] on the assessment and management of flood risks (Figure 10b).

The comparison results showed similarities even when the input material for their computation were at different scales and resolution. According to the National Administration "Romanian Waters", the hazard and risk maps for Bistriţa River were realized for the entire catchment at a spatial resolution between 5 and 10 m, while our analysis used a local 0.5 m input DEM where all the topographical details and hydraulic conditions were taken into consideration. Additionally, in a highly regulated river sector where several reservoirs are present, such as in our study area, the flood hazard is governed by their operating rules and management. Any mistake in the manipulation of water discharge can prove catastrophic for the areas downstream.

## 5. Conclusions

The 2D hydraulic modeling using HEC-RAS (v 5.0.3) and high-density LiDAR data applied for streamflow simulation within the urban and peri-urban area of Bacău City (North East Romania), produced sufficiently accurate information regarding flood hazard vulnerability. Based on multi-scenario development using different discharge releases from the rB1 (average discharge; calculated discharge: s1–s4) and its hydro-power plant (H-pp), the following concluding remarks can be summarized:

- Combining 2D hydraulic modeling with RS LiDAR data and local hydrological parameters proved to be an efficient method in order to improve the streamflow and spatial distribution of water over a densely urbanized and well-equipped hydro-technical area, in case of a flood event triggered by different discharges from the rB1 and its hydro-power plant (H-pp).
- The multi-scenario approach allowed the testing of flow capacity throughout the rB1 downstream floodplain, determining in this way, the discharge value which can cause a flood event. s1 and s2 scenarios kept the flood extent inside the river levees, meaning that a discharge up to 1160 $m^3/s$ for a period of one hour can be handled by the hydro-technical works downstream. The flood only starts to spread over the urban and peri-urban areas of Bacău City at a discharge rate over 1160 $m^3/s$ (s3 and s4).
- Scenario 2 (s2) allows a discharge of 6,636,960 $m^3$ of water, in a 24 h period (according to the input hydrograph) and captures the flood extent in the case where almost all the contained water in rB1 (Lilieci reservoir) is discharged. At a water surface elevation of 178.64 m and taking into consideration the existing LiDAR DEM, the reservoir stores 6,770,000 $m^3$ of water.

Overall, the 2D hydraulic modeling multi-scenario results can be exported into a set of flood hazard parameters such as flood depth, flood extent, flood velocity, or water surface elevation; and can answer real questions regarding the flood hazard threat at local level. Developing streamflow scenarios on a small-scale level is a very important aspect for any flood mitigation effort, especially in the urban areas located along main rivers, because large-scale analysis (river basin analysis) understates flood risk perception. Due to the adaptability of the 2D streamflow hydraulic model proposed in this study, the method can become a valuable asset in flood mitigation. However, a LiDAR DEM based 2D flood simulation is essential for every urbanized environment in the context of climate change and modern society development pressure and trends.

**Author Contributions:** Conceptualization, A.M.-P. and C.I.C.; methodology, A.M.-P., C.I.C. and M.N.P.; software, A.M.-P., C.I.C., C.C.S. and M.N.P.; validation, A.M.-P., C.I.C., C.C.S. and M.N.P.; formal analysis, A.M.-P. and C.I.C.; investigation, A.M.-P., C.I.C., C.C.S. and L.E.P.; resources, A.M.-P. and L.E.P.; data curation, A.M.-P. and C.I.C.; writing—original draft preparation, A.M.-P. and C.I.C.; writing—review and editing, A.M.-P.; supervision, A.M.-P.; project administration, A.M.-P.; funding acquisition, A.M.-P.

**Funding:** This work was funded by the Ministry of Research and Innovation of Romania within Program 1—Development of the national RD system, Subprogram 1.2—Institutional Performance—RDI excellence funding projects, Contract no.34PFE/19.10.2018 (beneficiary: A.M.-P.).

**Acknowledgments:** The authors would like to express their gratitude to the employees of the Romanian Waters Agency Bucharest, Siret Water Basin Administration Bacău, who kindly provided a significant part of the LiDAR data used in the present study. All data was processed in the Geoarchaeology Laboratory (Coordinator: A.M.-P.)

of Institute for Interdisciplinary Research, Science Research Department, "Alexandru Ioan Cuza" University of Iaşi (UAIC), Romania. Our thanks go to the all anonymous reviewers, who helped us in improving the manuscript.

**Conflicts of Interest:** The authors declare no conflict of interest. The founding sponsors had no role in the design of the study; in the collection, analyses, or interpretation of data; in the writing of the manuscript, and in the decision to publish the results.

## Abbreviations

The following abbreviations are used in this manuscript:

| | |
|---|---|
| DEM's | Digital Elevation Models |
| DSM | Digital Surface Models |
| EM-DAT | Emergency Events Database |
| EO | Earth Observation |
| GIS | Geographic Information System |
| HEC-RAS | Hydrologic Engineering Canter's—River Analysis System |
| H-pp | Hydro-power plant |
| LiDAR | Light Detection and Ranging |
| MLC | Maximum Likelihood Classification |
| MLIT | Ministry of Land Infrastructure and Transport (Japan) |
| NARW | National Administration "Romanian Waters" |
| rB1 | Bacău 1 reservoir (Lilieci storage reservoir) |
| rB2 | Bacău 2 reservoir |
| RS | Remote Sensing |
| s1–s4 | 2D streamflow hydraulic scenarios used in this study |
| SWBA | Siret Water Basin Administration |
| Sw1 | Bacău 1 dam spillway gates |
| Sw2 | Bacău 2 dam spillway gates |
| USACE | U.S. Army Corps of Engineers |
| USGS | US Geological Survey |

## References


1. Hall, J.; Arheimer, B.; Aronica, G.T.; Bilibashi, A.; Boháč, M.; Bonacci, O.; Borga, M.; Burlando, P.; Castellarin, A.; Chirico, G.B.; et al. A European Flood Database: Facilitating comprehensive flood research beyond administrative boundaries. *Proc. Int. Assoc. Hydrol. Sci.* **2015**, *370*, 89–95. [CrossRef]
2. Blöschl, G.; Hall, J.; Parajka, J.; Perdigão, R.A.P.; Merz, B.; Arheimer, B.; Aronica, G.T.; Bilibashi, A.; Bonacci, O.; Borga, M.; et al. Changing climate shifts timing of European floods. *Science* **2017**, *357*, 588–590. [CrossRef] [PubMed]
3. Alfieri, L.; Pappenberger, F.; Wetterhall, F.; Haiden, T.; Richardson, D.; Salamon, P. Evaluation of ensemble streamflow predictions in Europe. *J. Hydrol.* **2014**, *517*, 913–922. [CrossRef]
4. Jeneiová, K.; Kohnová, S.; Hall, J.; Parajka, J. Variability of seasonal floods in the Upper Danube River basin. *J. Hydrol. Hydromech.* **2016**, *64*, 357–366. [CrossRef]
5. Halgamuge, M.N.; Nirmalathas, A. Analysis of large flood events: Based on flood data during 1985–2016 in Australia and India. *Int. J. Disaster Risk Reduct.* **2017**, *24*, 1–11. [CrossRef]
6. Holmes, R.R., Jr.; Schwein, N.O.; Shadic, C.E. Flood Risk Awareness during the 2011 Floods in the Central United States: Showcasing the Importance of Hydrologic Data and Interagency Collaboration. *Leadersh. Manag. Eng.* **2012**, *12*, 101–110. [CrossRef]
7. Yoon, S.-K.; Kim, J.-S.; Moon, Y.-I. Integrated flood risk analysis in a changing climate: A case study from the Korean Han River Basin. *KSCE J. Civ. Eng.* **2014**, *18*, 1563–1571. [CrossRef]
8. Mertz, O.; Halsnæs, K.; Olesen, J.E.; Rasmussen, K. Adaptation to Climate Change in Developing Countries. *Environ. Manag.* **2009**, *45*, 743–752. [CrossRef] [PubMed]
9. Blöschl, G.; Montanari, A. Climate change impacts—throwing the dice? *Hydrol. Process.* **2010**, *24*, 374–381. [CrossRef]
10. Kvočka, D.; Falconer, R.A.; Bray, M. Appropriate model use for predicting elevations and inundation extent for extreme flood events. *Nat. Hazards* **2015**, *79*, 1791–1808. [CrossRef]

11. Sant'Anna, A.A. Not So Natural: Unequal Effects of Public Policies on the Occurrence of Disasters. *Ecol. Econ.* **2018**, *152*, 273–281. [CrossRef]
12. IPCC. Managing the Risks of Extreme Events and Disasters to Advance Climate Change Adaptation. In *A Special Report of Working Groups I and II of the Intergovernmental Panel on Climate Change*; Field, C.B., Barros, V., Stocker, T.F., Qin, D., Dokken, D.J., Ebi, K.L., Mastrandrea, M.D., Mach, K.J., Plattner, G.-K., Allen, S.K., et al., Eds.; Cambridge University Press: Cambridge, UK; New York, NY, USA, 2012; p. 582. Available online: https://www.ipcc.ch/site/assets/uploads/2018/03/SREX (accessed on 22 April 2019).
13. Bergholt, D.; Lujala, P. Climate-related natural disasters, economic growth, and armed civil conflict. *J. Peace Res.* **2012**, *49*, 147–162. [CrossRef]
14. Huang, L.; He, B.; Han, L.; Liu, J.; Wang, H.; Chen, Z. A global examination of the response of ecosystem water-use efficiency to drought based on MODIS data. *Sci. Total Environ.* **2017**, *601–602*, 1097–1107. [CrossRef] [PubMed]
15. Ahmadi, B.; Moradkhani, H. Revisiting Hydrological Drought Propagation and Recovery Considering Water Quantity and Quality. *Hydrol. Proc.* **2019**, *33*, 1492–1505. [CrossRef]
16. World Bank. Risk and Opportunity—Managing Risk for Development. In *World Development Report 2014*; License: Creative Commons Attribution CC BY 3.0; World Bank: Washington, DC, USA, 2014; p. 363. Available online: https://openknowledge.worldbank.org (accessed on 22 April 2019). [CrossRef]
17. Hoeppe, P. Trends in weather related disasters—Consequences for insurers and society. *Weather Clim. Extrem.* **2016**, *11*, 70–79. [CrossRef]
18. Banholzer, S.; Kossin, J.; Donner, S. The Impact of Climate Change on Natural Disasters. In *Reducing Disaster: Early Warning Systems for Climate Change*; Singh, A., Zommers, Z., Eds.; Springer: Dordrecht, The Netherlands, 2014; pp. 21–49.
19. Gigović, L.; Pamučar, D.; Bajić, Z.; Drobnjak, S. Application of GIS Interval Rough AHP Methodology for Flood Hazard Mapping in Urban Areas. *Water* **2017**, *9*, 360. [CrossRef]
20. Wilby, R.L.; Keenan, R. Adapting to flood risk under climate change. *Prog. Phys. Geogr. Earth Environ.* **2012**, *36*, 348–378. [CrossRef]
21. Wallemacq, P.; Guha-Sapir, D.; McClean, D.; CRED; UNISDR. *The Human Cost of Weather Related Disasters: 1995–2015*; Centre for Research on the Epidemiology of Disasters; UNISDR: Louvain, Belgium, 2015; p. 30. Available online: https://www.unisdr.org (accessed on 22 April 2019). [CrossRef]
22. Cojoc, G.M.; Romanescu, G.; Tirnovan, A. Exceptional floods on a developed river: Case study for the Bistrita River from the Eastern Carpathians (Romania). *Nat. Hazards* **2015**, *77*, 1421–1451. [CrossRef]
23. Romanescu, G.; Cimpianu, C.I.; Mihu-Pintilie, A.; Stoleriu, C.C. Historic flood events in NE Romania (post-1990). *J. Maps* **2017**, *13*, 787–798. [CrossRef]
24. Romanescu, G.; Nistor, I. The effect of the July 2005 catastrophic inundations in the Siret River's Lower Watershed, Romania. *Nat. Hazards* **2011**, *57*, 345–368. [CrossRef]
25. Romanescu, G.; Stoleriu, C.C. An inter-basin backwater overflow (the Buhai Brook and the Ezer reservoir on the Jijia River, Romania). *Hydrol. Process.* **2013**, *28*, 3118–3131. [CrossRef]
26. Romanescu, G.; Stoleriu, C.C. Causes and Effects of the Catastrophic Flooding on the Siret River (Romania) in July-August 2008. *Nat. Hazards* **2013**, *69*, 1351–1367. [CrossRef]
27. Romanescu, G.; Mihu-Pintilie, A.; Stoleriu, C.C.; Carboni, D.; Paveluc, L.; Cîmpianu, C.I. A Comparative Analysis of Exceptional Flood Events in the Context of Heavy Rains in the Summer of 2010: Siret Basin (NE Romania) Case Study. *Water* **2018**, *10*, 216. [CrossRef]
28. Bankoff, G. Constructing vulnerability: The historical, natural and social generation of flooding in metropolitan Manila. *Disasters* **2003**, *27*, 224–238. [CrossRef] [PubMed]
29. Dobrovičová, S.; Dobrovič, R.; Dobrovič, J. The Economic Impact of Floods and their Importance in Different Regions of the World with Emphasis on Europe. *Procedia Econ. Financ.* **2015**, *34*, 649–655. [CrossRef]
30. Cammerer, H.; Thieken, A.H.; Verburg, P.H. Spatio-temporal dynamics in the flood exposure due to land use changes in the Alpine Lech Valley in Tyrol (Austria). *Nat. Hazards* **2013**, *68*, 1243–1270. [CrossRef]
31. Mustafa, A.; Bruwier, M.; Archambeau, P.; Erpicum, S.; Pirotton, M.; Dewals, B.; Teller, J. Effects of spatial planning on future flood risks in urban environments. *J. Environ. Manag.* **2018**, *225*, 193–204. [CrossRef]
32. Mustafa, A.; Wei Zhang, X.; Aliaga, D.G.; Bruwier, M.; Nishida, G.; Dewals, B.; Erpicum, S.; Archambeau, P.; Pirotton, M.; Teller, J. Procedural generation of flood-sensitive urban layouts. *Environ. Plan. B Urban Anal. City Sci.* **2018**. [CrossRef]

33. Wheater, H.; Evans, E. Land use, water management and future flood risk. *Land Use Policy* **2009**, *26*, S251–S264. [CrossRef]
34. Donner, W.; Rodríguez, H. Population Composition, Migration and Inequality: The Influence of Demographic Changes on Disaster Risk and Vulnerability. *Soc. Forces* **2008**, *87*, 1089–1114. [CrossRef]
35. Kundzewicz, Z.W.; Kanae, S.; Seneviratne, S.I.; Handmer, J.; Nicholls, N.; Peduzzi, P.; Mechler, R.; Bouwer, L.M.; Arnell, N.; Mach, K.; et al. Flood risk and climate change: Global and regional perspectives. *Hydrol. Sci. J.* **2014**, *59*, 1–28. [CrossRef]
36. Vojtek, M.; Vojteková, J. Flood hazard and flood risk assessment at the local spatial scale: A case study. *Geomat. Nat. Haz. Risk* **2016**, *7*, 1973–1992. [CrossRef]
37. Boccard, N. Natural disasters over France a 35 years assessment. *Weather Clim. Extrem.* **2018**, *22*, 59–71. [CrossRef]
38. Foudi, S.; Osés-Eraso, N.; Tamayo, I. Integrated spatial flood risk assessment: The case of Zaragoza. *Land Use Policy* **2015**, *42*, 278–292. [CrossRef]
39. Teng, J.; Jakeman, A.J.; Vaze, J.; Croke, B.F.W.; Dutta, D.; Kim, S. Flood inundation modelling: A review of methods, recent advances and uncertainty analysis. *Environ. Model. Softw.* **2017**, *90*, 201–216. [CrossRef]
40. Balica, S.; Wright, N.G. A network of knowledge on applying an indicator-based methodology for minimizing flood vulnerability. *Hydrol. Process.* **2009**, *23*, 2983–2986. [CrossRef]
41. Samanta, S.; Pal, D.K.; Palsamanta, B. Flood susceptibility analysis through remote sensing, GIS and frequency ratio model. *Appl. Water. Sci.* **2018**, *8*, 66. [CrossRef]
42. Quiroga, V.M.; Kure, S.; Udo, K.; Mano, A. Application of 2D numerical simulation for the analysis of the February 2014 Bolivian Amazonia flood: Application of the new HEC-RAS version 5. *Ribagua* **2016**, *3*, 25–33. [CrossRef]
43. Armenakis, C.; Du, E.X.; Natesan, S.; Persad, R.A.; Zhang, Y. Flood Risk Assessment in Urban Areas Based on Spatial Analytics and Social Factors. *Geosciences* **2017**, *7*, 123. [CrossRef]
44. Pradhan, B.; Hagemann, U.; Tehrany, M.S.; Prechtel, N. An easy to use ArcMap based texture analysis program for extraction of flooded areas from TerraSAR-X satellite image. *Comput. Geosci.* **2014**, *63*, 34–43. [CrossRef]
45. Muthusamy, M.; Rivas Casado, M.; Salmoral, G.; Irvine, T.; Leinster, P. A Remote Sensing Based Integrated Approach to Quantify the Impact of Fluvial and Pluvial Flooding in an Urban Catchment. *Remote Sens.* **2019**, *11*, 577. [CrossRef]
46. Wan, W.; Liu, B.; Zeng, Z.; Chen, X.; Wu, G.; Xu, L.; Chen, X.; Hong, Y. Using CYGNSS Data to Monitor China's Flood Inundation during Typhoon and Extreme Precipitation Events in 2017. *Remote Sens.* **2019**, *11*, 854. [CrossRef]
47. Mihu-Pintilie, A.; Nicu, I.C. GIS-based Landform Classification of Eneolithic Archaeological Sites in the Plateau-plain Transition Zone (NE Romania): Habitation Practices vs. Flood Hazard Perception. *Remote Sens.* **2019**, *11*, 915. [CrossRef]
48. Zeleňáková, M.; Fijko, R.; Labant, S.; Weiss, E.; Markovič, G.; Weiss, R. Flood Risk Modelling of the Slatvinec Stream in Kružlov Village, Slovakia. *J. Clean. Prod.* **2019**, *212*, 109–118. [CrossRef]
49. Pasquier, U.; He, Y.; Hooton, S.; Goulden, M.; Hiscock, K.M. An integrated 1D–2D hydraulic modelling approach to assess the sensitivity of a coastal region to compound flooding hazard under climate change. *Nat. Hazards* **2018**, *94*, 1–23. [CrossRef]
50. Frank, E.; Ostan, A.; Coccato, M.; Stelling, G.S. Use of An Integrated One Dimensional-two Dimensional Hydraulic Modelling Approach for Flood Hazard and Risk Mapping. *WIT Trans. Ecol. Environ.* **2001**, *50*, 99–108. [CrossRef]
51. Patel, D.P.; Ramirez, J.A.; Srivastava, P.K.; Bray, M.; Han, D. Assessment of flood inundation mapping of Surat city by coupled 1D/2D hydrodynamic modeling: A case application of the new HEC-RAS 5. *Nat. Hazards* **2017**, *89*, 93–130. [CrossRef]
52. Schumann, G.J.P.; Neal, J.C.; Voisin, N.; Andreadis, K.M.; Pappenberger, F.; Phanthuwongpakdee, N.; Hall, A.C.; Bates, P.D. A first large-scale flood inundation forecasting model. *Water Resour. Res.* **2013**, *49*, 6248–6257. [CrossRef]
53. Bohorquez, P.; Darby, S.E. The use of one- and two-dimensional hydraulic modelling to reconstruct a glacial outburst flood in a steep Alpine valley. *J. Hydrol.* **2008**, *361*, 240–261. [CrossRef]

54. Sanders, B.F. Evaluation of on-line DEMs for flood inundation modeling. *Adv. Water Res.* **2007**, *30*, 1831–1843. [CrossRef]
55. Apel, H.; Aronica, G.T.; Kreibich, H.; Thieken, A.H. Flood risk analyses—how detailed do we need to be? *Nat. Hazards* **2009**, *49*, 79–98. [CrossRef]
56. Leandro, J.; Chen, A.S.; Djordjević, S.; Savic, D.A. Comparison of 1D/1D and 1D/2D Coupled (Sewer/Surface) Hydraulic Models for Urban Flood Simulation. *J. Hidraul. Eng.* **2009**, *135*, 495–504. [CrossRef]
57. Manfreda, S.; Samela, C.; Gioia, A.; Consoli, G.G.; Iacobellis, V.; Giuzio, L.; Cantisani, A.; Sole, A. Flood-prone areas assessment using linear binary classifiers based on flood maps obtained from 1D and 2D hydraulic models. *Nat. Hazards* **2015**, *79*, 735–754. [CrossRef]
58. Horritt, M.S.; Bates, P.D. Evaluation of 1D and 2D numerical models for predicting river flood inundation. *J. Hydrol.* **2002**, *268*, 87–99. [CrossRef]
59. Alho, P.; Aaltonen, J. Comparing a 1D hydraulic model with a 2D hydraulic model for the simulation of extreme glacial outburst floods. *Hydrol. Process.* **2008**, *22*, 1537–1547. [CrossRef]
60. Ghostine, R.; Hoteit, I.; Vazquez, J.; Terfous, A.; Ghenaim, A.; Mose, R. Comparison between a coupled 1D-2D model and a fully 2D model for supercritical flow simulation in crossroads. *J. Hydraul. Res.* **2014**, *53*, 274–281. [CrossRef]
61. Brovelli, M.A.; Cannata, M.; Longoni, U.M. LiDAR Data Filtering and DTM Interpolation Within GRASS. *Trans. GIS* **2004**, *8*, 155–174. [CrossRef]
62. Liu, X. Airborne LiDAR for DEM generation: Some critical issues. Progress in Physical Geography. *Earth Environ.* **2008**, *32*, 31–49. [CrossRef]
63. Garnero, G.; Godone, D. Comparisons between different interpolation techniques. *Int. Arch. Photogramm. Remote Sens. Spat. Inf. Sci.* **2013**, *XL-5/W3*, 139–144. [CrossRef]
64. Sibson, R. A Brief Description of Natural Neighbor Interpolation. In *Interpolating Multivariate Data*; Sibson, R., Ed.; John Wiley & Sons: New York, NY, USA, 1981; pp. 21–36.
65. OpenStreetMap. Available online: https://www.openstreetmap.org/ (accessed on 16 April 2019).
66. Betsholtz, A.; Nordlöf, B. Potentials and Limitations of 1D, 2D and Coupled 1D-2D Flood Modeling in HEC-RAS—A Case Study on Höje River. Master' Thesis, Division of Water Resources Engineering, Department of Building & Environmental Technology, Lund University, Lund, Sweden, 2017. Available online: http://lup.lub.lu.se/luur/download?func=download (accessed on 23 April 2019).
67. Smart, G.M. Improving flood hazard prediction models. *Int. J. River Basin Manag.* **2017**, *16*, 449–456. [CrossRef]
68. Alkema, D. Simulating Floods: On the Application of a 2D-hydraulic Model for Flood Hazard and Risk Assessment. Ph.D. Thesis, International Institute for Geo-information Science and Earth Observation, ITC Enschede, Enschede, The Netherlands, 2007. Available online: https://dspace.library.uu.nl/handle/1874/23283 (accessed on 23 April 2019).
69. Brunner, G.W. HEC-RAS 5.0 River Analysis System, Hydraulic Reference Manual. 2016. Available online: https://www.pdfdrive.com/hec-ras-hydraulic-reference-manual-e60353462.html (accessed on 23 April 2019).
70. Brunner, G.W.; CEIWR-HEC. HEC-RAS River Analysis System, 2D Modeling User's Manual; Version 5.0. 2016. Available online: https://www.pdfdrive.com/hec-ras-2d-modeling-users-manual-cpd-68a-e60355017.html (accessed on 23 April 2019).
71. Environment Agency. *What is the Updated Flood Map for Surface Water?* Environment Agency: Bristol, UK, 2013.
72. Bates, P.D.; De Roo, A. A simple raster-based model for flood inundation simulation. *J. Hydrol.* **2000**, *236*, 54–77. [CrossRef]
73. Vozinaki, A.-E.K.; Morianou, G.G.; Alexakis, D.D.; Tsanis, I.K. Comparing 1D and combined 1D/2D hydraulic simulations using high-resolution topographic data: A case study of the Koiliaris basin, Greece. *Hydrol. Sci. J.* **2017**, *62*, 642–656. [CrossRef]
74. Alzahrani, A.S. Application of Two-Dimensional Hydraulic Modeling in Riverine Systems Using HEC-RAS. Master Thesis, Master of Science in Civil Engineering, University of Dayton, Dayton, OH, USA, 2017. Available online: https://etd.ohiolink.edu/!etd.send_file?accession=dayton (accessed on 23 April 2019).
75. Gallegos, H.A.; Schubert, J.E.; Sanders, B.F. Two-dimensional, high-resolution modeling of urban dam-break flooding: A case study of Baldwin Hills, California. *Adv. Water Resour.* **2009**, *32*, 1323–1335. [CrossRef]
76. Earth Explorer. Available online: https://earthexplorer.usgs.gov/ (accessed on 16 April 2019).

77. Hu, S.; Tong, L.; Frazier, A.E.; Liu, Y. Urban boundary extraction and sprawl analysis using Landsat images: A case study in Wuhan, China. *Habitat Int.* **2015**, *47*, 183–195. [CrossRef]
78. INSPIRE Geoportal Access to European Spatial Data. Available online: http://inspire-geoportal.ec.europa.eu (accessed on 16 April 2019).
79. Cozma, D.G.; Cruceanu, A.; Cojoc, G.M.; Muntele, I.; Mihu-Pintilie, A. The factorial analysis of physic-chemical indicators in Bistrita's upper hydrographical basin. In Proceedings of the SGEM GeoConference 2015: SGEM 2015 GeoConference on Water Resources. Forest, Marine and Ocean Ecosystems, Albena, Bulgaria, 16–25 June 2015; Volume 15, pp. 625–632. [CrossRef]
80. Romanescu, G.; Mihu-Pintilie, A.; Ciurte, D.L.; Stoleriu, C.C.; Cojoc, G.M.; Tirnovan, A. Allocation of flood control capacity for a multireservoir system. Case study of the Bistriţa River (România). *Carpath. J. Earth Environ. Sci.* **2019**, *14*, 223–234. [CrossRef]

*water*

MDPI

*Article*

# How to Account for the Human Motion to Improve Flood Risk Assessment in Urban Areas

**Gabriele Bernardini** * and **Enrico Quagliarini**

Department of Construction, Civil Engineering and Architecture, Università Politecnica delle Marche, via Brecce Bianche, 60131 Ancona, Italy; e.quagliarini@staff.univpm.it

* Correspondence: g.bernardini@univpm.it; Tel.: +39-071-220-4246

Received: 23 March 2020; Accepted: 1 May 2020; Published: 7 May 2020

**Abstract:** Floods are critical disasters affecting urban areas and their users. Interactions with floodwater spreading and built environment features influence the users' reaction to the emergency, especially during immediate disaster phases (i.e., evacuation). Recent studies tried to define simulation models to evaluate such exposure-related criticalities, assess individuals' flood risk, and propose risk-mitigation strategies aimed at supporting the community's proper response. Although they generally include safety issues (e.g., human body stability), such tools usually adopt a simplified approach to individuals' motion representation in floodwaters, i.e., using input from non-specialized databases and models. This study provides general modelling approaches to estimate evacuation speed variations depending on individual's excitement (walking, running), floodwaters depths and individuals' features (age, gender, height, average speed on dry surfaces). The proposed models prefer a normalized evacuation speeds approach in respect of minimum motion constraint conditions to extend their applicability depending on the individuals' characteristics. Speed data from previous experiments are organized using linear regression models. Results confirm how individuals' speed reduces when depth and age increase. The most significant models are discussed to be implemented in evacuation simulation models to describe the evacuees' motion in floodwaters with different confidence degree levels and then assess the community's flood risk and risk-reduction strategies effectiveness.

**Keywords:** flood risk assessment; flood evacuation; evacuation modelling; behavioral design; urban built environment at risk; human motion in floodwaters

## 1. Introduction

Floods have been provoked over 93,000 victims worldwide in the years 2000–2018, being the second most disruptive natural disaster (after earthquakes) affecting our communities' safety (Source: EMDAT (2019): OFDA/CRED International Disaster Database, Université catholique de Louvain, Brussels, Belgium, https://www.emdat.be/, last access: 16 December 2019). Only in Italy, in the years 2014–2018, there were over 70 dead or missing persons, over 30 wounded, and over 21,000 individuals who had to evacuate and were homeless because of such kind of events (Source: http://polaris.irpi.cnr.it/report/last-report/, last access: 16 December 2019). In this general context, urban areas are the riskiest scenarios because of the combination between the built environment (in the following, BE) features (defined as a network of buildings, infrastructures and open spaces) and the high density of hosted exposed inhabitants [1–3]. Previous works demonstrated the significant importance of the interactions between the individuals and the flood-affected BE during the immediate response phases, and in particular during the emergency evacuation process [4–6]. In such conditions, the hosted population can attempt to move (drive or walk) in floodwaters-affected scenarios towards "safer" areas, with the aim to reduce threats [7–12]. Most of the flood-related fatalities occur in outdoor spaces during such activities and so it is essential to understand the underlying phenomena [13].

This is particularly evident in the following relevant scenarios [10,13–21]:

- Pre-emergency evacuation strategies cannot be implemented because of lack of early warnings or Civil Body Protection/First Responders management activities, thus making ineffective coordinated motor vehicles and public transport-based solutions (including, e.g., scenarios in underdeveloped countries);
- Flash flood conditions can exist;
- Flood-affected areas are pedestrian areas and/or the majority of the exposed population is placed outdoors, e.g., along the streets and the public squares;
- The motor vehicle evacuation cannot be performed because of specific conditions, e.g., (a) the urban space configuration (e.g., see compact urban fabric in historical scenarios); (b) critical effects in vehicle use along the evacuation roads, essentially due to vehicle-density related issues, such as traffic jams; (c) prevalence of pedestrian evacuees in respect to motorized ones (e.g., because of socio-economic factors, as for example, in underdeveloped countries);
- Distances to the evacuation areas (or the flood-affected area dimensions) are quite limited, such as in case of shelters in urban areas or "invacuation" strategies (According to PD 25111:2010, this evacuation solution concerns "the movement of people to pre-identified areas inside the building/site in order to protect them from external dangers during an incident". In flood evacuation, building occupants can remain inside and move towards the higher stories, while individuals placed in the building surrounding try to reach the nearest one and then move upstairs), and they should be reached on foot;
- Disruption of the mobility system can appear as leading people to move on foot;
- The real emergency scenario is quite different from what is expected, thus affecting one of the previous elements;
- Some First Responders activities can imply the movement on foot to support the population in the emergency scenarios.

The following cases can be combined and represent only an exemplification of the main conditions based on the literature review.

In such scenarios, individual's behaviors could lead to additional risks, too [6,10,22].

In this way, a brief overview on the current topics concerning a better risk assessment, counting for people moving in floodwaters and a proper effectiveness analysis of risk mitigation strategies, is reported in the next sub-sections, to have a general reference framework:

- Which are the risk-affecting factors in BE, by underlining the population-related contribute (Section 1.1);
- Which tools can be used to merge such two aspects to evaluate the effectiveness of the evacuation (Section 1.2);
- Which models can be used to evaluate man-floodwaters interactions in pedestrian evacuation (Section 1.3).

In particular, this study deals with this last issue, to improve the development of pedestrians' evacuation simulators in the context of flood disasters.

### 1.1. BE Risk-Affecting Factors and the Population-Related Contribute

Risk-increasing factors concerning the urban areas can be distinguished between those related to the BE itself (by including physical and management-related issues) and those related to the hosted population.

About the BE itself, they can be mainly related (but not limited) to [1,5,23–27]:

- Location of settlement on floodplains;

- Land use-related issues, by including a lower capacity of absorbing rain by urban surfaces (e.g., low porosity of urban surfaces, especially in high-built up areas) in respect to the ones in rural areas;
- Compact urban areas layout which can increase open-channel related effects in floodwater spreading;
- Insufficient sewer systems, because, e.g., of poor maintenance plans and actions as well as of the variations in functioning conditions (e.g., severity of floods different from the one used in their design) also due to climate-change effects;
- Vulnerability of buildings, suffering damage because of a low resistance level in respect to floodwater strains (e.g., elevations, foundations);
- Low level of control of possible hazard conditions, also in relation to poor flood sensor systems; inadequate management of immediate pre-disaster and disaster conditions, with the possible lack of early warning systems, which can provoke a delay in emergency action starting.

From the inhabitants' perspective, the general high population density in urban BE (in terms of inhabitants' number and localization on the urban layout) is matched with additional exposure-related factors such as social-economic factors, individuals' risk perception and awareness/preparedness and behavioral issues, which can lead people to additional dangers in emergency scenarios [5,10,19,28–30].

In such a framework, the two main elements characterizing the floodwaters and their direct effects on evacuees are their depth $D$ (m) and speed $V_f$ (m/s) [7–11]. In critical $D$-$V_f$ conditions, fatalities can be mainly due to: (a) interaction with debris dragged by floodwaters, with the possibility of minor injuries; (b) stability loss (e.g., for pedestrians, they can be due to possible buoyancy or body failure/dragging phenomena), with the consequent possibility of drowning and death. Similar problems can be related to both people moving on foot or by (motor) vehicles [7,8].

Besides, some human behavior in floods could increase risks during the evacuation process [4,6,10,13,17,29,31–33]:

- Emergency planning elements: poor awareness/preparedness levels of the population can bring people to adopt risky behaviors, underestimate the impact of disasters conditions, and delay the starting of safety procedures. Meanwhile, a similar result can be provoked by ineffective early warning systems. Emergency plans can be not well known by the population, hence making impossible to properly identify "safe" areas and gathering points, as well as evacuation paths;
- Surrounding BE elements: people generally try to move towards the nearest areas with lower floodwaters depth and speed (e.g., ground elevations), by looking for direct support of unmovable obstacles to move in floodwaters (e.g., fences, street furniture). Meanwhile, they try to modify their motion direction to avoid all the obstacles dragged by the flood (e.g., cars, debris);
- Other individuals and belongings: social share identity (including phenomena connected to information exchange) and attachment-to-things effects are time-wasting behaviors that can delay or slow down the evacuation process. Such elements can also bring people to move towards "unsafe" areas (e.g., to rescue other individuals, to retrieve some personal belongings, or to guard the properties). People can also prefer to move in a group by sharing evacuation direction and motion speed;
- Floodwaters: "curiosity" effects lead individuals to delay the evacuation starting to observe floodwaters conditions or event recording with smartphones or cameras. Meanwhile, when floodwaters depth and speed increase, the possibility to freely move is reduced and the evacuees' motion process is slowed down, hence increasing the overall evacuation time.

All these behaviors could lead people to interact with critical floodwater levels, by leading people to be exposed to more critical floodwater conditions because of, e.g., not being able to choose the proper evacuation direction or spend time in dangerous areas.

### *1.2. Simulation Tools to Risk-Assessment and Risk-Mitigation Strategies Evaluation*

Simulation models are effective evaluation tools for safety planners and designers for coordinating preparedness-oriented and response-oriented efforts of local Authorities and First Responders. Anyway, they should be developed by pursuing a behavioral point of view, to include the man-floodwater and the man-BE interactions as focus assessment elements [1,5,13,21,34].

As for other kinds of disasters [35–37], evacuation simulators could support this process by jointly representing behavioral issues, floodwater spreading and BE modifications over time and space. Most of the current simulators generally deal with a territorial/full urban scale, by mainly using motor vehicles or public transportation as evacuation means [5,15,26,38–40]. Such kind of model could introduce simplifications in the evacuation rules by adopting "macroscopic" models (e.g., density-related models for motion estimation). This choice is reasonable if considering the widescale application but can limit the simulator effectiveness while it locally evaluates threats for the population, especially in case of local interactions with the floodwaters (i.e., for pedestrian evacuation). Meanwhile, the number of simulators dealing with pedestrian evacuation seems to be still limited [13,21,34]. In view of the above, "microscopic" approaches should be preferred [10,34,41]. These approaches assign motion rules to each of the simulated individuals to derive the overall evacuation phenomena and then estimate the safety level for the population [21]. Hence, they allow evaluating risks at the typical scale (dimension and configuration) of each element composing the BE (i.e., streets and other open spaces, buildings) [13].

### 1.3. Modelling Pedestrians' Evacuation in Floodwater: Current Approaches and Limitations

The "microscopic" approach defines the overall motion phenomena by overlapping the effects of different behavioral rules assigned to each simulated evacuee [10,41]. To this aim, it is important to analyze the behavioral rules of an "isolated" pedestrian, that is, an individual moving into the floodwater by himself/herself (free-flowing movement conditions). Additional interaction behaviors could be then overlapped to this, to define the overall motion rules. Furthermore, individuals' variations in behavior could be randomly assigned, by allowing to describe different evacuees' attitude in the evacuation process. In this view, besides the definition of qualitative behavioral rules in flood evacuation [6,10,17], the determination of motion quantities plays a pivotal role for the following reasons.

Previous studies demonstrate how differences between general-purpose and flood-related motion quantities exist, especially about the evacuation speed, which is affected by the levels of the floodwaters [10,33,42,43]. Efforts to assess the individual's evacuation speed $v_i$ (m/s) have been recently performed through real-world event analysis [10] or laboratory experiments in open channels [33,44] or pools [42]. Nevertheless, only a limited number of works succeeds in overlaying the main constraints due to [33,36,42]: (a) the limited number of individuals, also in relation to the different age classes; (b) the effects of individuals' excitement conditions (i.e., walking versus running).

This is the object of the present paper: $D$ is here considered as the main floodwater evacuation affecting parameter in both walking and running conditions, as well as, its correlations with individuals' gender and age.

## 2. Phases, Materials, and Methods

This work is organized into three main phases:

1. General characterization of the selected database, according to Section 2.2 methods. The motion speeds in walking and running conditions are compared depending on $D$ and on the individuals' gender. In this way, it can be possible to respectively trace general relationships depending on floodwater-characterization constraints to motion and to individuals' main features;
2. Modelling of the speed variation (normalized speed and dimensional speed variations) depending on $D$ and in respect of the dry surface motion, according to Section 2.3 methods. In particular, dry surface motion speeds are retrieved according to average literature values depending on the age [43]. The individuals' height (i.e., the knee height from the ground) is considered in the model because of the related influence assessed by preliminary works on motion and human body stability conditions in floodwaters [9,44];

3. Modelling of the normalized speed variation depending on $D$ in respect to the minimum constraint and maximum excitement conditions for the considered database, according to methods in Section 2.4. Such model delves into specific experimental conditions and sample features, to generalize speed estimation data regardless of dry surface motion conditions data.

### 2.1. Materials: Original Input Database Characterization

The database by Bernardini et al. [44] is used as the input reference for the model definition (Raw data are available as supplementary files). More than 200 individuals were involved in laboratory experiments by moving in walking and running excitement conditions along a 24-m-long path into an open channel (rectangular section of 1 × 1 m; concrete horizontal bottom surface; individuals wearing fishermen's suits). A total of 555 samples compose the overall database.

The volunteers' sample was defined according to the Italian National age statistics and included individuals from 8 to 83 years old (males: 56%, females: 44%). Tested floodwaters $D$ were equal to 0.2, 0.3, 0.4, 0.5, 0.6 and 0.7 m, by considering still water ($V_f = 0$ m/s) to better focus on the effects of depth on $v_i$. For each individual involved in the test, the original database includes: (a) the motion speeds $v_i$ (m/s) referred to the specific tested floodwaters and excitement conditions; (b) the individuals' features (i.e., sex, gender, height, mass, body mass index); (c) the number of performed tests. Each individual performed the test once in walking and running conditions, for a given $D$, and 31 individuals performed the test for all the $D$ values.

### 2.2. General Methods for Model Definition and Characterization of the Selected Database

A preliminary analysis to assess the relative position of the knee height in respect of the floodwater depth has been performed, to evidence if the original database could be affected by the individuals' height $H$ (m). In fact, previous works on human body kinematics suggest that differences in individuals' motion could exist for intermediate $D$ values (e.g., 0.4 to 0.6 cm) in case the knee is outside or inside the water level. In particular, a physiological constraint (i.e., "the difficulty to articulate the motion of the legs") can be induced while moving the knee and the part of the above leg inside the floodwaters [44]. In general terms, the knee height $K$ (m) can be calculated as $0.285 \times H$ [44,45]. The distribution of $K$-$D$ ((cm) or (m)) in respect to $D$ tested values is assessed and shown by a boxplot comparison, while linear regression is provided to evidence the general trend of the sample.

Then, the database has been mainly characterized by the comparison between walking and running conditions through $v_{w,exp}$-$v_{r,exp}$ (m/s) pairs analysis depending on:

- $D$, to evaluate the impact of floodwaters constraint conditions. It could be assumed that higher $D$ implies smaller differences between $v_{w,exp}$ and $v_{r,exp}$ [33,44];
- Gender, to evaluate difference among the two excitement conditions and compare them with previous works outcomes (i.e., males generally moves faster than females) [43].

Linear regression models are tested by reporting the $R^2$ values to focus on the direct correlation between the two assessed variables. Then, simple models to be integrated into simulators can be traced. According to previous works on the original database [44], all the models have been evaluated through the Bisquare regression method, which allows finding the regression that fits the bulk of the data and minimizes the effect of outliers.

For each of this analysis, the $v_{w,ex} = v_{r,exp}$ line (no differences between walking and running conditions) is shown to confirm if the above regressions are over this ratio.

### 2.3. Methods for Speed Variation Modelling Depending on D and Dry Surface Motion Conditions

The analysis methods described in this section are applied to the whole original sample, regardless of the number of tests performed by the volunteers. The database significant dimension is similar to those of previous experimental activities chosen as reference works [43,46].

Previous studies tried to derive correlations between the individuals' speeds $v_i$ (m/s) and $D$ in absolute terms (m/s) [33,34,44], but only a limited number of works tried to define such correlations in respect of free-flowing and dry surface motion conditions [33], that is, the no-constraint conditions for human body motion also in evacuation conditions. The use of such normalized speed could instead improve the generalization of simulation models by allowing to overlap normalized speed variations and the main motion-affecting individuals' parameters (e.g., age, gender) [43,46]. Thus, this second approach has been followed, and the individuals' speeds $v_i$ (m/s) have been expressed in relation to maximum or ideal evacuation speeds and so in non-dimensional terms (-).

The input data are then organized to trace the variation of speed in walking and running conditions depending on the dry surface motion conditions according to literature values [43]. For each database element, the average motion speed on dry surfaces (m/s) is calculated as a function of the related individuals' age, according to the correlation curves given by previous studies on wide databases (see the curves in Figure 1). Hence, two values of dry surfaces average motion speed depending on age are provided for each individual: $v_{id,a}$, according to the curve $a$ [46] (see Figure 1a) and $v_{id,b}$, according to the curve $b$ [43] (see Figure 1b). These two curves offer different age-related values, and in particular, the curve $b$ [43] can be selected to consider precautionary conditions in normal motion since it generally offers speed values lower than the one of the curve $a$ [46].

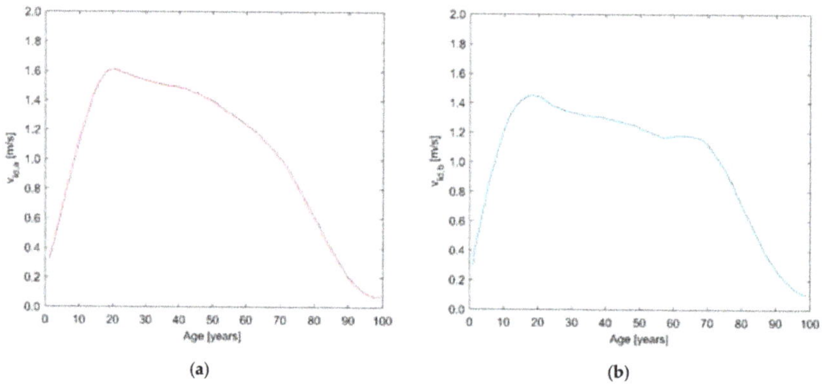


**Figure 1.** Reference curves for speed normalization depending on the individuals' age according to: (**a**) the curve $a$, to derive $v_{id,a}$ [46] and (**b**) to the curve $b$, to derive $v_{id,b}$ [43].

Then, Equation (1) is applied to derive the individual's normalized speed $v^*$ (-) according to the two considered curves, in running ($v^*_{a,r}$ and $v^*_{b,r}$) and walking ($v^*_{a,w}$ and $v^*_{b,w}$) conditions:

$$v^*_{a,r} = v_{r,exp}/v_{id,a},\ v^*_{b,r} = v_{r,exp}/v_{id,b},\ v^*_{a,w} = v_{w,exp}/v_{id,a},\ v^*_{b,w} = v_{w,exp}/v_{id,b} \quad (1)$$

These values are then organized according to the following modelling approaches, to retrieve according to linear regression models for the prediction of the output in both walking and running conditions:

1. Variation of the normalized speed $v^*$ in respect of $D$ experimental classes: the distribution of $v^*$ in respect of $D$ experimental classes takes advantages of boxplot analysis and graph to evidence if and how the input data are widespread and if outliers are present. Input data involve the whole sample and male (outlined by subscript $M$) and female (outlined by subscript $F$) separate samples to highlight the differences due to gender. In general terms, it is expected that running regression model should involve higher $v^*$ values, especially for lower $D$-related input data;
2. Variation of the normalized speed $v^*$ in respect of $K$-$D$: linear regression models are tested. Such modelling approach allows detecting the effects of the reciprocal position of knee and floodwater

levels. Male and female samples are shown in the regression model but considered together in the regression model to focus on the main factor influencing human behavior, which is the individuals' height. It is expected that $v^*$ should be equal to or higher than 1 when the knee is placed outside the floodwaters, especially in running conditions.

In addition, a dimensional age-evacuation speed (m/s) model has been tested for the different $D$ values according to a polynomial (4th degree) regression model, which considered the general literature reference curves trend [43,46]. This model allows tracing the differences between the motion curves in no-constraint conditions versus $D$-constraints conditions. Male and female samples are considered together in the regression model. This choice allows: (a) focusing on the main factor influencing the human behavior, which is the individuals' age and (b) comparing this study curves with the ones of reference works. The regression models are tested for the whole sample (average behavior of the tested population) and for the maximum values of the experimental pairs (maximum limit behavior). Meanwhile, minimum values are not considered due to their similarities regardless of the specific $D$ values, as demonstrated in the original input database assessment [44]. It is assumed that lower $D$ curves could produce speed values higher than the references curves, especially for running conditions.

### 2.4. Methods for Normalized Speed Variation Modelling Depending on D, Minimum Constraint, and Maximum Excitement Conditions for the Considered Database

The analysis reported in this section concerns the database subsample including the individuals who performed all the experiments for each $D$. It includes 186 tests performed by 31 individuals (age characterization: mean age of 32 years, standard deviation of 9 years, range from 21 to 66 years; height characterization: mean height of 1.71 m, standard deviation of 0.07 m, range from 1.55 m to 1.83 m; 16 females and 15 males). The analysis of such data is considered to trace the variation of speed in walking and running conditions depending on the maximum excitement conditions (running) with minimum floodwater experimental constraint ($D$ = 0.2 m). In fact, it is possible to compare the motion behaviors of considered volunteers for each of the tested conditions.

For each individual, the ideal normalized speed with respect to the maximum excitement-minimum experimental constraint conditions (-) is calculated as in Equation (2):

$$v_{i,r,0.2} = v_{r,exp}/v_{r,0.2},\ v_{i,w,0.2} = v_{w,exp}/v_{r,0.2} \tag{2}$$

where $v_{r,0.2}$ (m/s) is the individuals' running speed for $D$ = 0.2 m. For this reason, all the $v_{i,r,0.2}$ for $D$ = 0.2 m will be equal to 1. The distribution of $v_{i,r,0.2}$ (cm) in respect to $D$ tested values is assessed and shown by a boxplot comparison. Linear regression is provided to evidence the general trend of the sample for separated running and walking excitement conditions. Due to the smaller sample dimension, it is assumed that the regressions are evaluated by collecting males and females in a unique sample.

In addition, the linear regressions are also tested for the sub-sample of individuals who moved with the knee inside and outside the floodwater, according to the same procedure. To this end, Equation (3) has been applied to separately retrieve the normalized speed value in running $v_{i,r,max}$ and walking $v_{i,w,max}$ excitement conditions:

$$v_{i,r,max} = v_{r,exp}/v_{r,ma},\ v_{i,w,max} = v_{w,exp}/v_{w,max} \tag{3}$$

where $v_{r,max}$ and $v_{w,max}$ are respectively the maximum running and maximum walking speed. This normalization procedure allows to consider the effective $D$ value for which the individual had the maximum motion speed. Hence, the boxplot distribution representation can have $v_{i,r,max}$ and $v_{i,w,max}$ values equal to 1 also for $D$ > 0.2 m.

## 3. Results

### 3.1. Characterization of the Selected Database

Figure 2 demonstrates that the overall sample can be characterized by a significant linear trend in *K-D* (cm) values. The main classes affected by the possibility to have the knee inside or outside of the floodwater levels are $D = 0.4$ m and 0.5 m. Hence, the results for these classes can be effectively affected by the relative position between the knee and the floodwater surface. Lower *D* classes have many outliers with respect to the others because of the values connected to the child, which have a lower height.

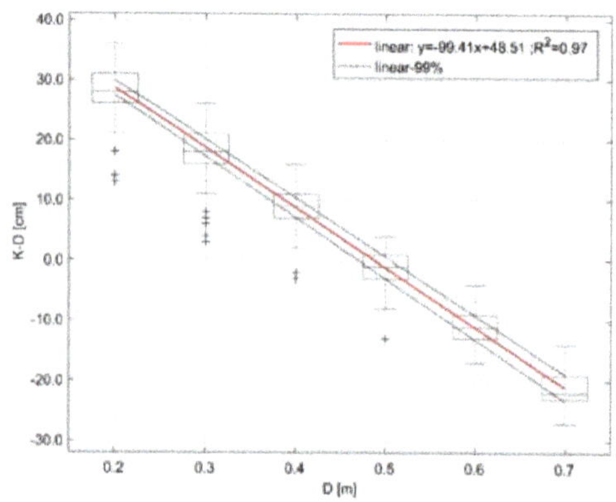


**Figure 2.** Database characterization depending on *K-D* versus the tested *D* values. The boxplot graph includes possible outliers (shown by "+"). The linear regression and the 99% boundaries are shown.

Figure 3 confirms how the male sample individuals seem to move faster in running conditions than the female sample individuals [43]. Anyway, differences between the two linear regression models are slight (about 10% in respect of the male regression coefficient). $R^2$ values show a moderate correlation trend for both the gender-related sample.

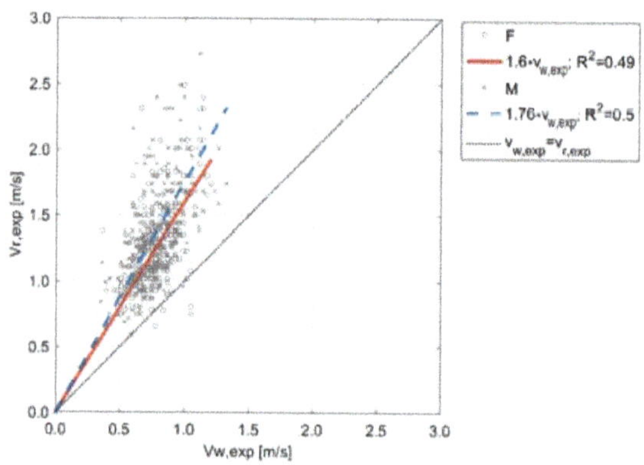


**Figure 3.** Database characterization depending on $v_{w,exp}$ and $v_{r,exp}$ pairs, for each individual, by distinguishing females (continuous red line) and males (dashed blue line) linear regressions and related samples.

Figure 4 confirms how the increase of *D* implies a slighter difference between running and walking excitement conditions, with a general moderate correlation trend (compare $R^2$ values for the regression models). $D$ equal to 0.4 m and 0.5 m evidences small differences in regression trends (5% in respect of $D = 0.5$ m) because the related sample is composed by individuals with knees outside and inside the floodwater level, also according to Figure 2 results. Meanwhile, $D$ equal to 0.5 m and 0.6 m seems to additionally provoke the same conditions in $v_{w,exp}$ and $v_{r,exp}$ pairs, confirming previous works outcomes [44]. In such conditions, the knee and a part of the above leg are generally placed inside the water, by involving a similar physiological and kinematic constraint in human body motion.

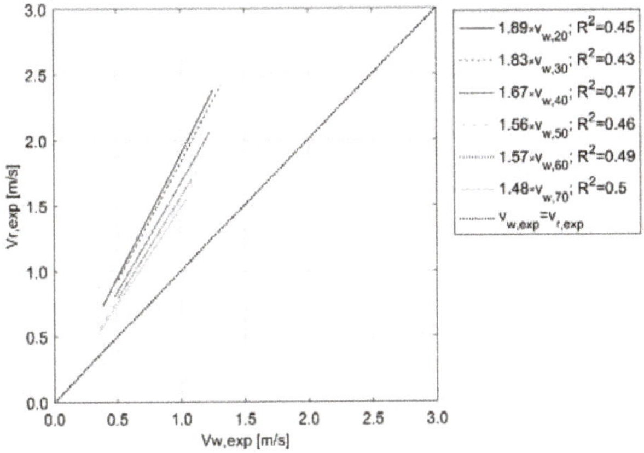


**Figure 4.** Database characterization depending on $v_{w,exp}$ and $v_{r,exp}$ pairs, for each individual, by distinguishing the linear regressions for each of the tested $D$. The regression equations are shown as a function of the respective $v_{w,exp}$ for the considered $D$ value (subscript expressed in (cm)).

### 3.2. Speed Variation Modelling Depending on D and Dry Surface Motion Conditions

Figure 5 resumes the normalized speed variation models according to the two reference curves, for running conditions, while Figure 6 evidences the same model in walking conditions. In general terms, all the provided models confirm how [33,41,43,44]:

1. Motion speed (and so $v^*$) decreases while $D$ increases. Regressions on the curve $a$-based model show a slighter regression slope, according to the first term in the regression equation (−8% for running conditions and −16% for walking conditions in respect to the curve $a$-based model). Hence, curve $b$-based models seem to amplify the $v^*$ reduction;
2. The speed-$D$ relationship in both the excitement conditions can be properly represented by using speed normalization. In particular, regressions on the curve $a$ [46]-based model show a moderate relationship, according to the $R^2$ value, for both walking and running conditions. $R^2$ values for the curve $b$ [43]-based model are lower than these, suggesting that this model more limitedly represents the effects of $D$ in $v^*$ variation. Hence, the adoption of curve $a$-based model could improve the prediction model capabilities in evacuation simulators;
3. In view of the above, differences between running and walking excitement conditions exist. In particular, the effects of $D$ on the variation of speed in walking conditions seem to be less relevant than the ones of running conditions. This result is evidenced by the more significant presence of possible outliers (see Figure 6) and by the lower value of the first coefficient of the linear regression (about −70% in respect to running conditions model). This result confirms the original work outcomes [44] and the outcomes of studies concerning other evacuation typologies [47]. In fact, issues related to the individual's motion effort and the distance between individual and motion goals seem to be more relevant while people are running towards the motion goal (as in evacuation conditions). In addition, it is worth noting that the assessed $v^*$ value in the dry surface for running conditions is about 2 times greater than the one for walking conditions, regardless of

the adopted model. Besides, $v^*$ in running conditions seems to be equal to about 1 when $D$ is about 0.3 to 0.5m, which corresponds to the $D$ classes for which the individuals' knee is generally closer to the floodwater level, according to Figure 2 statistics;

4. Differences between dry surface motion in literature reference works [43,46] and the adopted database [44] exist (compare the second linear regression coefficient in all the models), by underlining how specific database should be adopted to represent the flood evacuation. Forecasted $v^*$ for $D < 0.2$ m are beyond the lower limit of the tested $D$ conditions and so they could be affected by additional non-linear speed-$D$ interferences [33]. Nevertheless, the importance of this result is shown by the $v^*$ values for the minimum tested constraint conditions ($D = 0.2$ m) in both walking and running conditions.

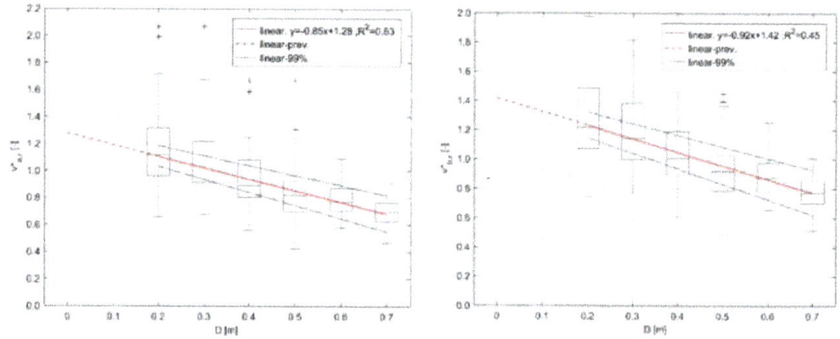


(a) (b)

**Figure 5.** Normalized speed variation for running conditions depending on $D$ in respect to dry surface motion conditions, for the whole sample (male and female together): (**a**) according to the curve $a$ [46]; (**b**) according to the curve $b$ [43]. Linear regressions (dashed lines refer to prevised values outside of the tested $D$ range) and boundary conditions (99%) are shown. Outliers are shown in the boxplot representation (marked by "+").

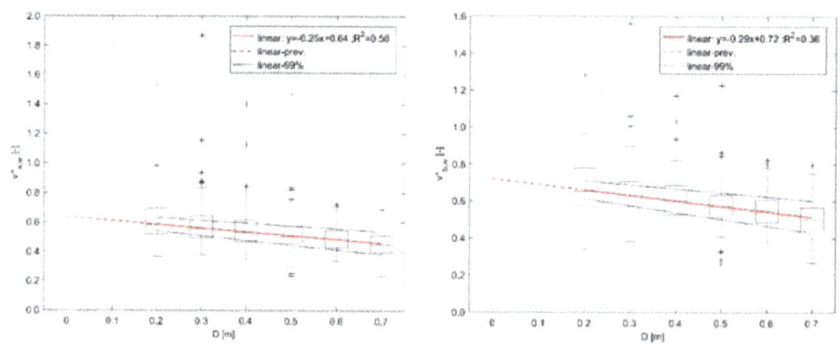


(a) (b)

**Figure 6.** Normalized speed variation for walking conditions depending on $D$ in respect to dry surface motion conditions, for the whole sample (male and female together): (**a**) according to the curve $a$ [46]; (**b**) according to the curve $b$ [43]. Linear regressions (dashed lines refer to prevised values outside of the tested $D$ range) and boundary conditions (99%) are shown. Outliers are shown in the boxplot representation (marked by "+").

The same results are highlighted by the running and walking models related to male and female separated samples, as shown by Figures 7 and 8. Besides the higher accuracy of the curve $a$-based model in prediction $v^*$ variation depending on $D$, the comparison between running and walking conditions for male and female subsamples underlines how:

- Differences in walking conditions for the two subsamples in respect to the general subsample are really limited (compare the regression coefficients of Figure 6a to Figures 7c and 8c, and the regression coefficients of Figure 6b to Figures 7d and 8d);
- Males seem to move faster than females in running conditions, confirming previous works outcomes [33,43] and Section 3.1 (i.e., compare with Figure 3) outcomes. This result is evidenced by the second linear regression coefficient in both the models (in curve *a*-based model: 1.33 for males versus 1.23 for females; in curve *b*-based model: 1.47 for males versus 1.36 for females). In particular, males seem to move +8% faster (in normalized terms) in dry surface conditions, in both models;
- At the same time, the increase of $D$ has the same effect on both males and females, in both models, since the first regression coefficient does not significantly change (compare Figure 7a to Figure 8a and Figure 7b to Figure 8b).

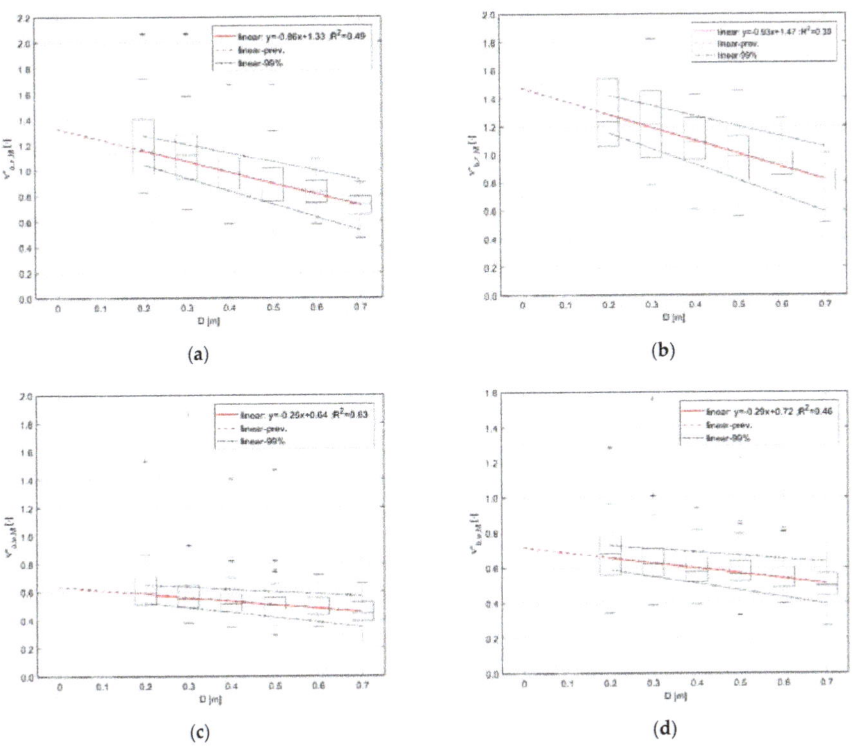

**Figure 7.** Normalized speed variation depending on $D$ in respect to dry surface motion conditions, for the male sample: (**a**) for running conditions, according to the curve *a* [46]; (**b**) for running conditions, according to the curve *b* [43]; (**c**) for walking conditions, according to the curve *a* [46]; (**d**) for walking conditions, according to the curve *b* [43]. Linear regressions (dashed lines refer to expected values outside of the tested $D$ range) and boundary conditions (99%) are shown. Outliers are shown in the boxplot representation (marked by "+").

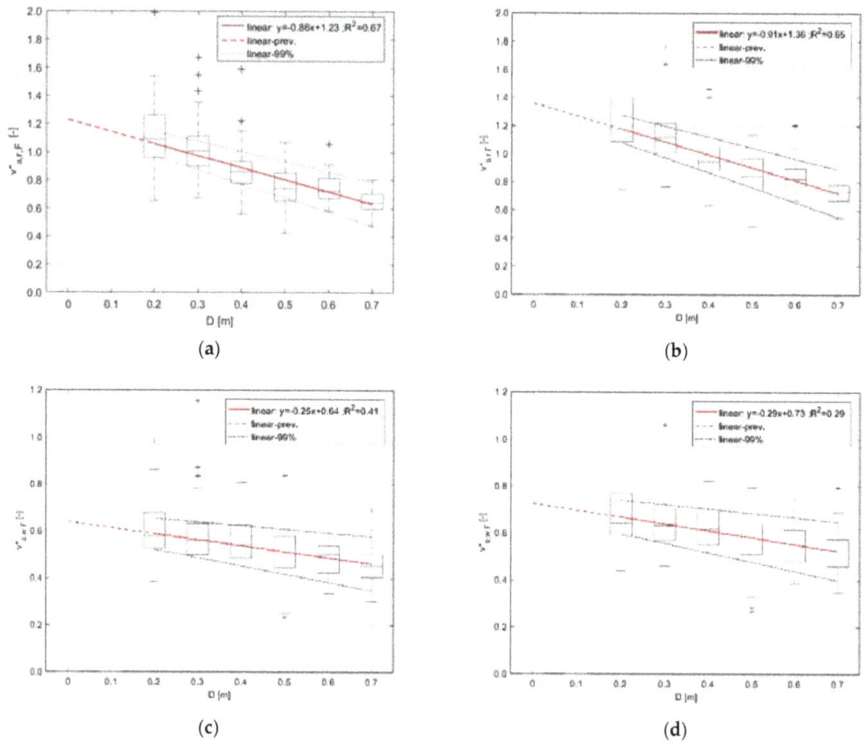


**Figure 8.** Normalized speed variation depending on $D$ in respect to dry surface motion conditions, for the female sample: (**a**) for running conditions, according to the curve $a$ [46]; (**b**) for running conditions, according to the curve $b$ [43]; (**c**) for walking conditions, according to the curve $a$ [46]; (**d**) for walking conditions, according to the curve $b$ [43]. Linear regressions (dashed lines refer to expected values outside of the tested $D$ range) and boundary conditions (99%) are shown. Outliers are shown in the boxplot representation (marked by "+").

Figures 9 and 10 confirm the previous results by focusing on the knee-floodwater depth interaction and by representing the normalized speed variation according to the reciprocal knee-$D$ position. The database characterization according to $K$-$D$ values is offered by previous Figure 2. The male and female samples are jointly considered to focus the regression model on the individuals' height-related behavioral driver. Firstly, according to Figures 6 and 7 outcomes, the effect of $D$ on the human motion is more relevant in running conditions (see the first regression coefficient in all the models). Secondly, $v^*$ is quite equal to 1 for $K$-$D$ values close to 0 in both running conditions related models. Finally, all the regression models (both using curve $a$ and curve $b$) seem to have a more relevant statistical significance in respect of the $v^*$-$D$ ones ($R^2$ generally higher also for curve $b$-based model). All these results confirm the importance of floodwater-knee interaction as the main behavioral driver for the whole sample [44].

It is also worth noting that the general dispersion of forecaster $v^*$ data is quite small and increases for individuals moving with the knee outside of the water (i.e., $K$-$D > 0$ m). This result is shown by the reciprocal position of the linear boundary conditions (99%) in respect of the linear regression line (which refers to forecasted median values).

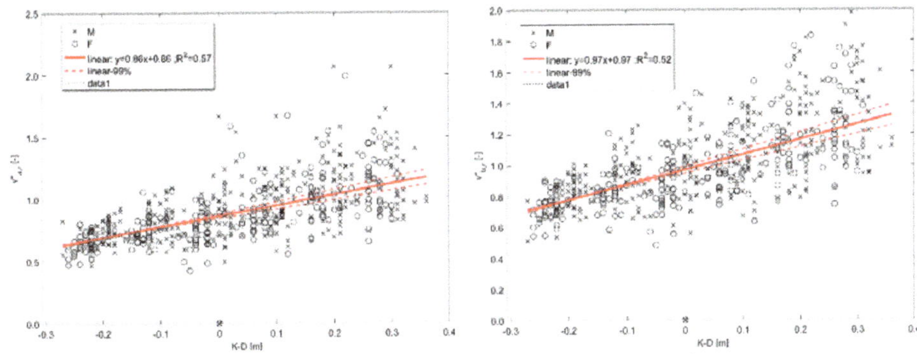


(a) (b)

**Figure 9.** Normalized speed variation for running conditions depending on *K-D* in respect to dry surface motion conditions, for the whole sample (male and female together): (**a**) according to the curve *a* [46]; (**b**) according to the curve *b* [43]. Linear regressions (dashed lines refer to expected values outside of the tested *D* range) and boundary conditions (99%) are shown. Male and female samples experimental pairs are shown.

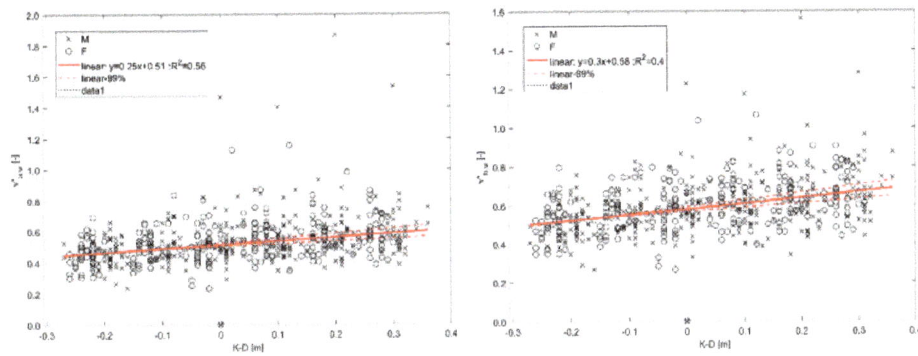


(a) (b)

**Figure 10.** Normalized speed variation for walking conditions depending on *K-D* in respect to dry surface motion conditions, for the whole sample (male and female together): (**a**) according to the curve *a* [46]; (**b**) according to the curve *b* [43]. Linear regressions (dashed lines refer to expected values outside of the tested *D* range) and boundary conditions (99%) are shown. Male and female samples experimental pairs are shown.

The *age*-evacuation speed *v* (m/s) models are assessed at the different tested *D* values for the whole sample (average regression) and maximum *age-v* pairs, by using a 4th-degree polynomial regression which has a similar form in respect of the reference curves. Figures 11 and 12 respectively show the results for the overall sample in running and walking conditions, and by considering the comparison between the curve *a* and the curve *b*. Figures 13 and 14 show the regression for the maximum *age-v* pairs. In each excitement condition, considered age being equal, the increase of *D* corresponds to a decrease in predicted motion speed. Differences are more relevant in running tests. The regression models could be affected by the original database characterization in terms of age classes and overall age range [44], as outlined by low $R^2$ values (<0.25) for "average" *age-v* regression line (due to the original database pairs dispersion). Nevertheless, they generally confirm the previous modelling outcomes as well as the results of previous works. Firstly, the general trend of 4th polynomial regression is confirmed in each excitement conditions, by underlining that elderly's speeds are lower than the one of adult individuals. In addition, it could be evidenced that:

- Considering running conditions, the *age-v* regression for the overall sample is over the references curves when $D$ ranges from 0.2 m to 0.4 m, regardless of the considered curve in the model; this result confirms the "average" results of Figure 9. Meanwhile, the regression curves on the maximum *age-v* pairs seem to be always over the reference curves or, at least, equal (i.e., if considering the curve $b$ data);
- On the contrary, walking conditions-related regression on maximum *age-v* pairs are generally under the reference curves, with similar predicted values for the age range 10 to 30 years. This seems to evidence that younger individuals can generally be less affected in their motion by $D$ conditions, confirming previous results on age-related impact on motion speed [42–44]. Average *age-v* pairs regression in walking speeds seems to evidence no substantial differences, especially for intermediate ranges, confirming the interpretation of outcomes in Figures 6 and 10.

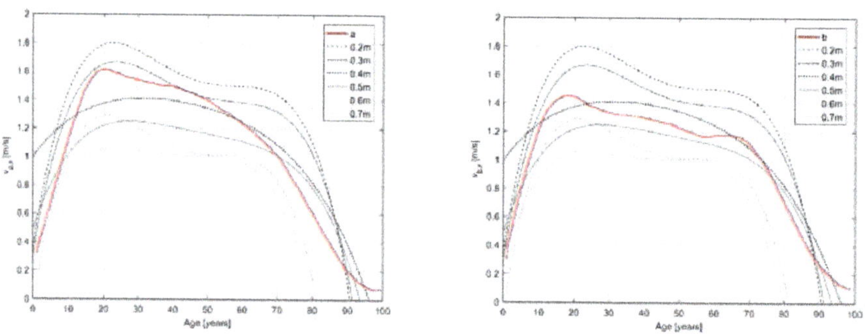


(a) (b)

**Figure 11.** Average dimensional age-evacuation speed (m/s) model in running conditions, for the different $D$ values according to a polynomial (4th degree) regression model, performed on all the experimental pairs (male and female together): (**a**) compared to the curve $a$ [46]; (**b**) compared to the curve $b$ [43].

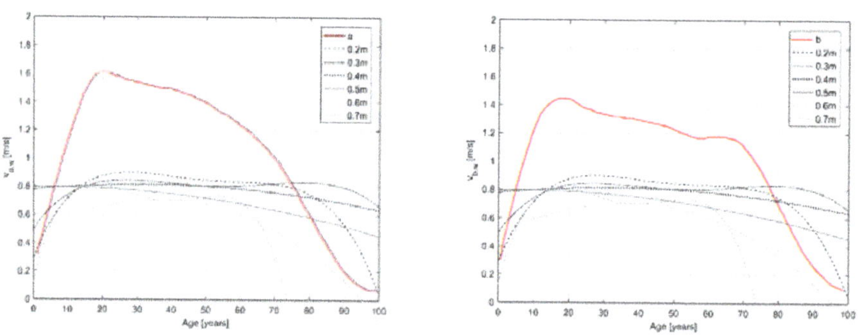


(a) (b)

**Figure 12.** Average dimensional age-evacuation speed (m/s) model in walking conditions, for the different $D$ values according to a polynomial (4th degree) regression model, performed on all the experimental pairs (male and female together): (**a**) compared to the curve $a$ [46]; (**b**) compared to the curve $b$ [43].

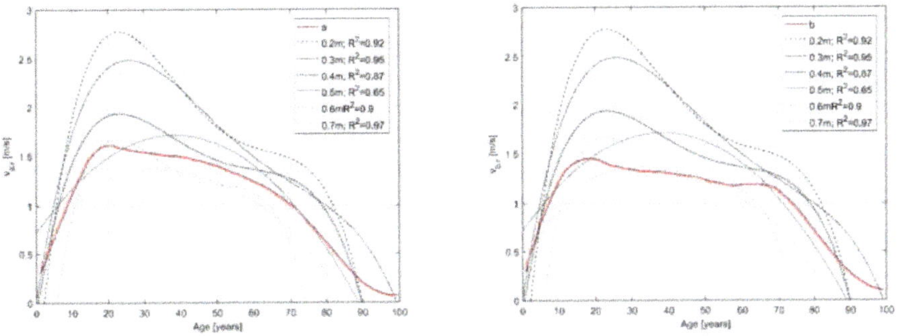


**(a)** **(b)**

**Figure 13.** Maximum dimensional age-evacuation speed (m/s) model in running conditions, for the different $D$ values according to a polynomial (4th degree) regression model, performed on all the experimental pairs (male and female together): (**a**) compared to the curve *a* [46]; (**b**) compared to the curve *b* [43].

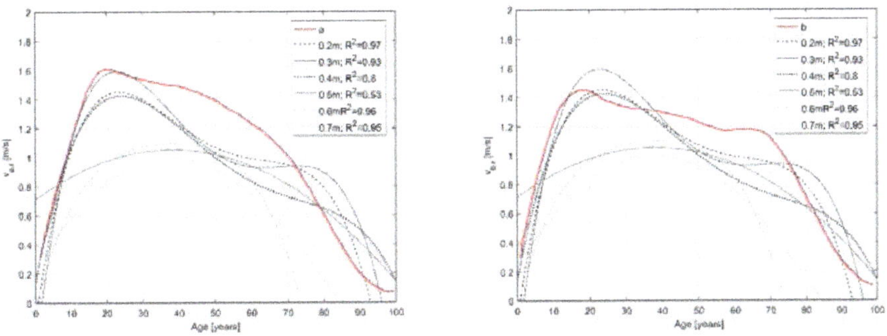


(a) (b)

**Figure 14.** Average dimensional age-evacuation speed (m/s) model in walking conditions, for the different $D$ values according to a polynomial (4$^{th}$ degree) regression model, performed on all the experimental pairs (male and female together): (**a**) compared to the curve *a* [46]; (**b**) compared to the curve *b* [43].

### *3.3. Normalized Speed Variation Modelling Depending on D, Minimum Constraint, and Maximum Excitement Conditions for the Considered Database*

Figure 15 traces the normalized speed trends in respect to the minimum constraint-maximum excitement conditions ($D$ = 0.2 m; running experiments), depending on $D$, for data concerning running (Figure 15a) and walking (Figure 15b) tests. Although the sample dimension is quite limited (31 individuals for 186 considered tests), the results confirm previous trends discussed in Section 3.2 concerning the speed reduction as a function of $D$ increase.

In particular, Figure 15a evidences how the maximum motion speed is higher for lower $D$ values in running conditions. Boxplot distribution data for $D$ = 0.2 m are connected to the normalization of experimental speed by $v_{r,0.2}$ (hence, all the values are equal to 1), but the maximum $v_{i,r,0.2}$ values decrease while $D$ increases. In addition, the median linear regression denotes a moderate relationship between $D$ and $v_{i,r,0.2}$, as shown by the $R^2$ value. On the contrary, walking conditions-related speeds $v_{i,w,0.2}$ seems to be less influenced by $D$, as demonstrated by the scattered boxplot values distribution and by the inconsistent $R^2$ value.

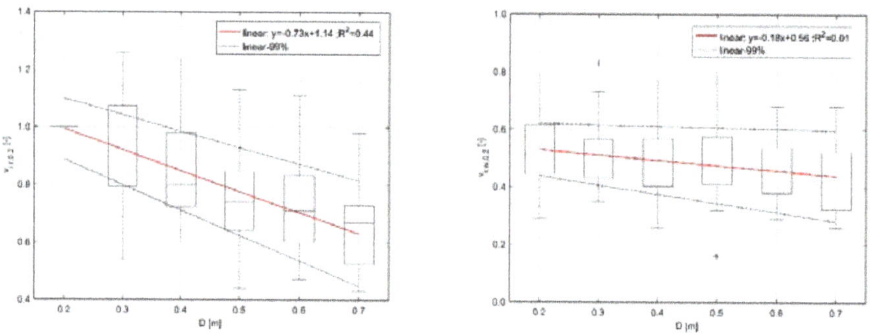


(**a**) (**b**)

**Figure 15.** Normalized speed variation depending on *D*, by normalizing the speed values by $v_{r,0.2}$ (speed of the individual in minimum constraint and maximum excitement conditions within the tested *D* range): (**a**) for running conditions; (**b**) for walking conditions. Linear regressions (dashed lines refer to prevised values outside of the tested *D* range) and boundary conditions (99%) are shown.

Finally, Figures 16 and 17 examine the effect of Knee position in respect to the floodwater depth, respectively in running and walking conditions, by considering the normalization according to the individuals' effective maximum speed within the tested *D* range. Although the regressions generally show a lower statistical significance according to poorer $R^2$ values, it is demonstrated that:

In running conditions, most of $v_{i,r,max}$ values are linked to the lowest D values, while walking conditions see a more widespread $v_{i,w,max}$ distribution. This outcome evidences the existence of excitement-related issues (i.e., motion effort increasing with constraints increase) in pseudo-evacuation excitement conditions, in respect to normal excitement conditions (pseudo-evacuation conditions can be approximated by running test, while normal conditions by walking tests) [44,47].

- The effect of *D* on an individual's speed is higher while the pedestrian is moving with the knee outside of the floodwater level, especially in running conditions as shown for the sub-sample with the knee inside the floodwater by: (a) the lower $R^2$ value, (b) the wider range for boundary conditions (99%) regression lines, and (c) the lower first regression coefficient value (only for running conditions of Figure 16).

Such outcomes verify the results reported in Section 3.2 about the normalized speed variation modelling depending on *D* and the dry surface motion condition. Meanwhile, they use data on the experimental *D* range from individuals who performed all the tests (possibility to trace the individuals' response in all the considered scenarios).

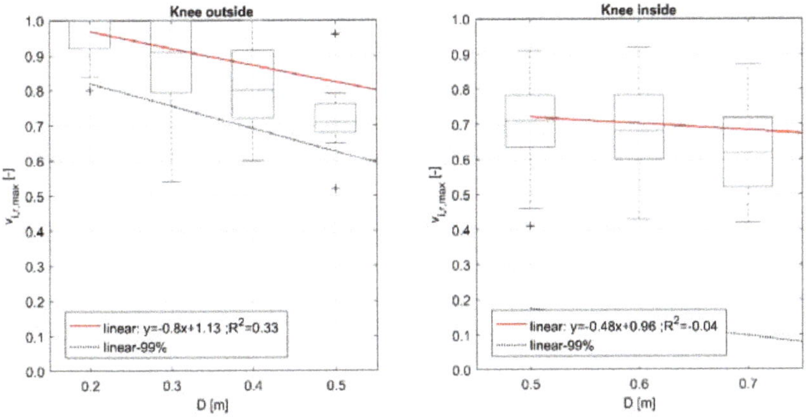


**Figure 16.** Effects of knee position (outside and inside the floodwater level) on the normalized speed variation depending on $D$, by normalizing the speed values by $v_{r,max}$. The regressions are shown for running conditions. Linear regressions (dashed lines refer to prevised values outside of the tested $D$ range) and boundary conditions (99%) are shown.

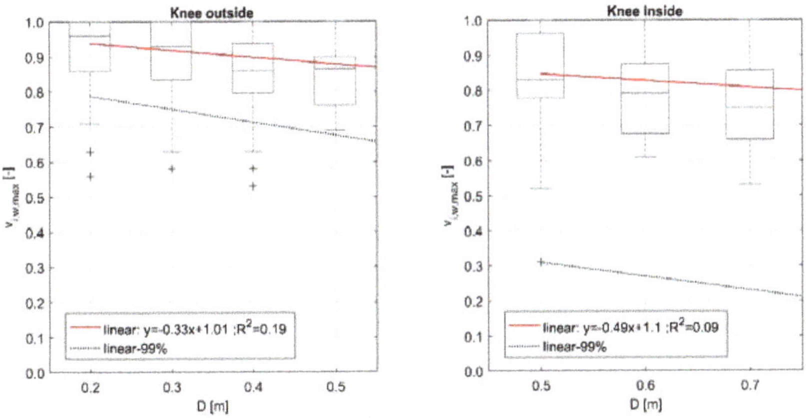


**Figure 17.** Effects of knee position (outside and inside the floodwater level) on the normalized speed variation depending on $D$, by normalizing the speed values by $v_{w,max}$. The regressions are shown for walking conditions. Linear regressions (dashed lines refer to prevised values outside of the tested $D$ range) and boundary conditions (99%) are shown.

## 4. Results Discussion in View of Model Implementation in Flood Evacuation Simulators

The modelling approaches offered by this work represent different solutions for the implementation of quantitative behavioral response aspects (i.e., individuals' speed) in flood evacuation simulators.

Firstly, all of them are based on a microscopic point of view and so they can be applied in such kind of simulation models [10,34,39,48]. According to this microscopic standpoint, they trace the "isolated" pedestrian speed by considering a single environmental driver, which is the floodwater depth and his/her specific features (mainly age, gender, and excitement level). In some terms, it could be defined as the ideal speed at which the evacuees try to tend. Anyway, in a real environment, this ideal speed could be affected by some other behavioral drivers [10,41], such as those described in Section 1.1. Hence, the final motion velocity (considering the speed (that is, the velocity magnitude), its direction, and its verse) of each individual should be calculated by overlapping all these phenomena. In this context, it is important to underline that some specific effects can provide consistent modifications to the effective motion velocity:

- Effects due to the presence of other pedestrians: crowd-density effects could reduce the speeds in compact groups; social shared identity factors could lead people to share the same direction and to adjust their individual speeds to remain in group (e.g., waiting for more vulnerable and slow group members);
- Guidance and support elements: they can speed up the process by giving continuous support to the evacuee in terms of direction to be followed and in terms of physical aid (e.g., handrails along the paths);
- Other flood-related phenomena like the attraction towards unmovable obstacles, according to the aforementioned support needs of the evacuees.

In view of the above, the proposed microscopic standpoint allows increasing the possibility of applying the proposed model to different modelling techniques (e.g., force-driven, velocity-driven, agent-based).

Secondly, most of the proposed modelling approaches also include specific individuals' features, like age, gender and height. Such choice allows considering the possible variation in such parameters within the simulated population, especially while adopting agent-based modelling techniques [21,34].

Thirdly, the modelling approaches focus on the two different excitement conditions for pedestrians [41,43,44]:

- Input running conditions used in this study are similar to the one of excited individual, who, for instance, participate to an evacuation process with a high engagement level (e.g., because of rising floodwater levels or other hazard-increasing evacuation drivers perceived by the individuals);
- Input walking conditions used in this study are similar to the one of normal motion individual, who for instance, participate in an evacuation process in still waters, or under organized circumstances.

In this way, simulators could adopt the related models depending on the assumed excitement conditions (e.g., walking conditions imply the maximization of motion time, moving towards conservative evaluations). At the same time, simulations under different excitement levels could be performed to assess the alterations among them [41].

Finally, in view of the above, Tables 1 and 2 summarize the selected modelling approaches (according to the model significance in terms of fitting performance and description of evacuation phenomena) that can be included in flood evacuation simulators, by distinguishing between:

- Dimensional models, which are the ones in which the dependent variables are directly expressed in dimensional terms (i.e., the dependent variable is a speed (m/s) (Table 1). These models allow tracing the simplest solutions to predict motion speed in flood evacuation according to a microscopic approach, by directly retrieving evacuation speed values in dimensional terms (m/s) only according to a single motion driver [41,43];
- Normalized models, which are the ones in which the dependent variables are a non-dimensional parameter $v_{norm}$ (-) (Table 2). To calculate the effective motion speed, $v_{norm}$ should be multiplied by the isolated individuals' speed, which can be the one in (a) minimum constraint-maximum excitement conditions in the tested database ($v_{r,0.2}$) or (b) dry surface motion ($v_{id}$). Although the validity of the model application can be extended to different isolated pedestrians' speed databases, the use of reference curves by this work for models development is encouraged especially in the second aforementioned case (i.e., [46], and so to $v_{id,a}$, due to the higher statistical significance of results).

Finally, the proposed models focus on different drivers of human behaviors by confirming the existence of the main driver in individuals' response while moving in floodwaters [6,9,10,33,44].

Nevertheless, it is worth noting that their statistical significance is strictly connected to the original database adopted by this work.

**Table 1.** Selected modelling approaches based on dimensional dependent variables.

| Independent Variables (unit of measure) | Dependent Variable (unit of measure) | Regression Model (sample) [1,2] | $R^2$ [1] | Specific Discussion |
|---|---|---|---|---|
| walking speed $v_{w,exp}$ (m/s); gender (F or M) | running speed $v_{r,exp}$ (m/s) | $v_{r,exp} = 1.6v_{w,exp}$ (F)<br>$v_{r,exp} = 1.76v_{w,exp}$ (M) | ≈0.5 | Figure 3, Section 3.1 |
| walking speed $v_{w,exp}$ (m/s); floodwater depth $D$ (m) | running speed $v_{r,exp}$ (m/s) | $D = 0.2$ m: $v_{r,exp} = 1.89v_{w,exp}$ (A)<br>$D = 0.3$ m: $v_{r,exp} = 1.83v_{w,exp}$ (A)<br>$D = 0.4$ m: $v_{r,exp} = 1.67v_{w,exp}$ (A)<br>$D = 0.5$ m, 0.6 m: $v_{r,exp} = 1.56v_{w,exp}$ (A)<br>$D = 0.7$ m: $v_{r,exp} = 1.48v_{w,exp}$ (A) | ≈0.5 | Figure 4, Section 3.1 |
| age (years) | "maximum" running speed $v_{r,exp}$ (m/s) | 4th polynomial regression model graphically traced in Figure 13 | >0.7 | Figure 13, Section 3.2 |
| age (years) | "maximum" walking speed $v_{r,exp}$ (m/s) | 4th polynomial regression model graphically traced in Figure 14 | >0.7 | Figure 14, Section 3.2 |

[1] Coefficients are valid for the considered sample; further activities should be performed to check their validity. [2] F refers to female sample, M refers to male sample, A refers to both male and female samples together.

**Table 2.** Selected modelling approaches based on non-dimensional dependent variables. In the regression model column, the expression inside round brackets is $v_{norm}$.

| Independent Variables (Unit of Measure) | Dependent Variable (Unit of Measure) | $v_{norm}$ Regression Model (Sample) [1,2] | $R^2$ [1] | Specific Discussion |
|---|---|---|---|---|
| floodwater depth D (m) | running speed $v_r$ (m/s) | $v_r = (-0.85D + 1.28)v_{id,a}$ (A) | ≈0.5 | Figure 5a, Section 3.2 |
| floodwater depth D (m) | walking speed $v_w$ (m/s) | $v_w = (-0.25D + 0.64)v_{id,a}$ (A) | ≈0.6 | Figure 6a, Section 3.2 |
| floodwater depth D (m); gender (F or M) | running speed $v_r$ (m/s) | $v_r = (-0.86D + 1.23)v_{id,a}$ (F)<br>$v_r = (-0.86D + 1.33)v_{id,a}$ (F) | >0.5 | Figures 7a and 8a, Section 3.2 |
| relative position between the knee height and the floodwater level K-D (m) | running speed $v_r$ (m/s) | $v_r = (-0.86D + 0.86)v_{id,a}$ (A) | ≈0.6 | Figure 9a, Section 3.2 |
| relative position between the knee height and the floodwater level K-D (m) | walking speed $v_w$ (m/s) | $v_w = (-0.25D + 0.51)v_{id,a}$ (A) | ≈0.6 | Figure 10a, Section 3.2 |
| floodwater depth D (m) | running speed [3] $v_r$ (m/s) | $v_r = (-0.73D + 1.14)v_{r,0.2}$ (A) | ≈0.4 | Figure 15a, Section 3.3 |

[1] Coefficients are valid for the considered sample; further activities should be performed to check their validity. [2] F refers to female sample, M refers to male sample, A refers to both male and female samples together. [3] This model considers the original database-related speed and not the reference curve as the multiplication value of $v_{norm}$.

## 5. Conclusions and Future Research Remarks

The flood risk assessments in urban areas should consider the evaluation of hosted community's exposure to possible emergency conditions, especially if considering the possible man-floodwaters interactions that play a pivotal role in immediate emergency scenarios, i.e., in the evacuation process. The effectiveness verification of emergency planning and other risk-mitigation strategies implemented in the built environment (BE) and oriented at direct support of the involved population should take advantage of simulation-based analysis.

This work offers a fundamental step toward this goal by providing general and unified modelling approaches to represent the individuals' evacuation speed (m/s) depending on (a) his/her walking speeds (m/s) and gender, (b) the floodwater depth (m), (c) his/her age (years), and (d) the relative position between the knee height and the floodwater level (m). According to previous research outcomes, these models confirm how the individuals' speed decreases when age and floodwater depths increase, while males generally seem to move faster than females. Moreover, they are generally characterized by moderate or strong relationship confidence degree, guaranteeing their applicability to modelling purposes. Finally, walking speeds can be evaluated by using inverse correlation modelling based on the proposed ones concerning running speed tests.

Although the used database is significant, different databases could be selected to collect input data and extend the validity of proposed regression models or compare/validate them. The considered original database describes characterization only in terms of floodwater depth. Hence, future research should try including additional significant variables, such as floodwater speed, sediment transport/mud presence. Additional experimental activities are needed to provide significant data to this end.

Besides, the current work is structured by defining regression models by preferring a continuous data existence field (for input and outputs). Nevertheless, widest databases could support the development of multiple linear models to derive the relationship between the moving speed and other factors (e.g., age, flood depth) one by one. Coherence models (e.g., regression tree-based) could be provided to have complete modelling result in a microscopic approach.

The proposed modelling approaches could be introduced in flood evacuation simulators by providing input data, which are more accurate than one of the general-purpose existing databases. Microscopic-based simulation tools can take advantage of this work's outcomes since they assign evacuation quantities to each simulated individual and allow considering specific individuals' features (i.e., age, gender, excitement level). The combination of behavioral simulation with flood spreading simulation in the BE could allow assessing the evacuation timing and so if/how many individuals could be exposed to significant risk depending on the floodwater hazard evaluation over space. Such simulators could additionally include rules for motor vehicle and public transport-based evacuation over time, to evidence which dependencies among effects could emerge about evacuation typologies and pedestrian-vehicle interactions. Finally, they could also support the analysis of First Responders' motion in flood-affected scenarios during the emergency response phases.

From this point of view, risk assessment analyses for different flood events and in different BE layouts will be achieved to trace the effectiveness of risk-reduction interventions. Proposed risk reduction strategies could involve, e.g., early warning systems, interventions on buildings and architectural components in public spaces, emergency planning and management including the possibility to schedule (delay, anticipate, divide per areas) the evacuation starting, interventions for controlling floodwater spreading, and evacuees' wayfinding systems. Furthermore, they could be combined with each other. Simulation-based approaches including the evacuation process will allow evaluating how such public efforts could effectively support efficiency gains also in case of risky conditions directly affecting the population and provoking the necessity to evacuate. The positioning over space of some solutions could be also evaluated.

Meanwhile, they will evaluate which are the minimum strategies that could be introduced in the BE, e.g., by moving towards minimum conditions in: (a) implementation costs (to reduce public efforts by local authorities), (b) management of complexity (to reduce the organizational efforts requested by the First Responders); (c) impact on the BE features (i.e., if considering historical urban heritage). In this sense, such kind of evaluation could be of interest especially in those urban scenarios where pedestrians could be caught in the middle of the flood "by surprise" (e.g., due to possible limitations in warning and communication with the population) as well as where difficulties due to social and economic factors could increase the overall urban risk (e.g., in underdeveloped countries).

**Supplementary Materials:** The following are available online at http://www.mdpi.com/2073-4441/12/5/1316/s1, manuscript-supplementary.xlsx: raw data (Excel file) used as the input reference for the model definition [44].

**Author Contributions:** Conceptualization, E.Q.; methodology, G.B. and E.Q.; validation, G.B. and E.Q.; formal analysis, G.B.; investigation, G.B.; resources, E.Q. and G.B.; data curation, G.B.; writing—original draft preparation, G.B.; writing—review and editing, E.Q. and G.B.; visualization, G.B.; supervision, E.Q.; project administration, E.Q.; funding acquisition, E.Q. All authors have read and agreed to the published version of the manuscript.

**Funding:** This research was funded by the scientific project "Building Resilience to Flood Impact Deriving from Global Warming in Europe (BRIDGE)" funded by the Polytechnic University of Marche internal program 2017/2018.

**Conflicts of Interest:** The authors declare no conflict of interest.

## References


1. Paquier, A.; Mignot, E.; Bazin, P.-H. From Hydraulic Modelling to Urban Flood Risk. *Procedia Eng.* **2015**, *115*, 37–44. [CrossRef]
2. Kontokosta, C.E.; Malik, A. The Resilience to Emergencies and Disasters Index: Applying big data to benchmark and validate neighborhood resilience capacity. *Sustainable Cities Soc.* **2018**, *36*, 272–285. [CrossRef]
3. Chen, K.F.; Leandro, J. A Conceptual Time-Varying Flood Resilience Index for Urban Areas: Munich City. *Water* **2019**, *11*, 830. [CrossRef]
4. Hamilton, K.; Peden, A.E.; Pearson, M.; Hagger, M.S. Stop there's water on the road! Identifying key beliefs guiding people's willingness to drive through flooded waterways. *Saf. Sci.* **2016**, *89*, 308–314. [CrossRef]
5. Jamrussri, S.; Toda, Y. Available Flood Evacuation Time for High-Risk Areas in the Middle Reach of Chao Phraya River Basin. *Water* **2018**, *10*, 1871. [CrossRef]
6. Chanson, H.; Brown, R.; McIntosh, D. Human body stability in floodwaters: The 2011 flood in Brisbane CBD. In Proceedings of the Hydraulic Structures and Society—Engineering Challenges and Extremes: Proceedings of the 5th IAHR International Symposium on Hydraulic Structures (ISHS2014), Brisbane, Australia, 25–27 June 2014; The University of Queensland: Australia; pp. 1–9.
7. Xia, J.; Falconer, R.A.; Lin, B.; Tan, G. Numerical assessment of flood hazard risk to people and vehicles in flash floods. *Environ. Modell. Softw.* **2011**, *26*, 987–998. [CrossRef]
8. Milanesi, L.; Pilotti, M.; Ranzi, R. A conceptual model of people's vulnerability to floods. *Water Resour. Res.* **2015**, *51*, 182–197. [CrossRef]
9. Cox, R.J.; Shand, T.D.; Blacka, M.J. *Australian Rainfall and Runoff Revision Project 10: Appropriate Safety Criteria for People*; Engineers Australia—Water Engineering: Barton, ATC, Australia, 2010; ISBN 9780858259454. Available online: http://arr.ga.gov.au/__data/assets/pdf_file/0006/40578/ARR_Project_10_Stage1_report_Final.pdf (accessed on 4 May 2020).
10. Bernardini, G.; Camilli, S.; Quagliarini, E.; D'Orazio, M. Flooding risk in existing urban environment: From human behavioral patterns to a microscopic simulation model. *Energy Procedia* **2017**, *134*, 131–140. [CrossRef]
11. Hu, P.; Zhang, Q.; Shi, P.; Chen, B.; Fang, J. Flood-induced mortality across the globe: Spatiotemporal pattern and influencing factors. *Sci. Total Environ.* **2018**, *643*, 171–182. [CrossRef]
12. Confederation of Fire Protection Associations Europe. *European Guideline—CFPA-E No 1:2012 N Protection against Flood Damage*; CFPA Europe: Copehangen, Denmark, 2012. Available online: http://www.cfpa-e.eu/wp-content/uploads/files/guidelines/CFPA_E_Guideline_No_1_2012_N.pdf (accessed on 4 May 2020).
13. Arrighi, C.; Pregnolato, M.; Dawson, R.J.; Castelli, F. Preparedness against mobility disruption by floods. *Sci. Total Environ.* **2019**, *654*, 1010–1022. [CrossRef]
14. Melo, R.; Zêzere, J.L.; Oliveira, S.C.; Garcia, R.A.C.; Oliveira, S.; Pereira, S.; Piedade, A.; Santos, P.P.; van Asch, T.W.J. Defining evacuation travel times and safety areas in a debris flow hazard scenario. *Sci. Total Environ.* **2020**, *712*, 136452. [CrossRef] [PubMed]
15. Wood, N.; Jones, J.; Peters, J.; Richards, K. Pedestrian evacuation modeling to reduce vehicle use for distant tsunami evacuations in Hawai'i. *Int. J. Disaster Risk Reduction* **2018**, *28*, 271–283. [CrossRef]
16. Lin, J.; Zhu, R.; Li, N.; Becerik-Gerber, B. How occupants respond to building emergencies: A systematic review of behavioral characteristics and behavioral theories. *Saf. Sci.* **2020**, *122*, 104540. [CrossRef]
17. Takagi, H.; Li, S.; de Leon, M.; Esteban, M.; Mikami, T.; Matsumaru, R.; Shibayama, T.; Nakamura, R. Storm surge and evacuation in urban areas during the peak of a storm. *Coastal Eng.* **2016**, *108*, 1–9. [CrossRef]
18. Kolen, B.; van Gelder, P.H.A.J.M. Risk-Based Decision-Making for Evacuation in Case of Imminent Threat of Flooding. *Water* **2018**, *10*, 1429. [CrossRef]
19. Kawasaki, A.; Kawamura, G.; Zin, W.W. A local level relationship between floods and poverty: A case in Myanmar. *Int. J. Disaster Risk Reduction* **2020**, *42*, 101348. [CrossRef]
20. Fan, Q.; Tian, Z.; Wang, W. Study on Risk Assessment and Early Warning of Flood-Affected Areas when a Dam Break Occurs in a Mountain River. *Water* **2018**, *10*, 1369. [CrossRef]
21. Bernardini, G.; Postacchini, M.; Quagliarini, E.; Brocchini, M.; Cianca, C.; D'Orazio, M. A preliminary combined simulation tool for the risk assessment of pedestrians' flood-induced evacuation. *Environ. Modell. Softw.* **2017**, *96*, 14–29. [CrossRef]

22. Ishigaki, T.; Toda, K.; Baba, Y.; Inoue, K.; Nakagawa, H.; Yoshida, Y.; Tagawa, H. Experimental study on evacuation from underground space by using real size models. *Proc. Hydraul. Eng.* **2006**, *50*, 583–588. [CrossRef]
23. Wu, J.; Ye, M.; Wang, X.; Koks, E. Building Asset Value Mapping in Support of Flood Risk Assessments: A Case Study of Shanghai, China. *Sustainability* **2019**, *11*, 971. [CrossRef]
24. Morelli, S.; Segoni, S.; Manzo, G.; Ermini, L.; Catani, F. Urban planning, flood risk and public policy: The case of the Arno River, Firenze, Italy. *Appl. Geogr.* **2012**, *34*, 205–218. [CrossRef]
25. Milman, A.; Warner, B.P. The interfaces of public and private adaptation: Lessons from flooding in the Deerfield River Watershed. *Glob. Environ. Chang.* **2016**, *36*, 46–55. [CrossRef]
26. Hissel, F.; Morel, G.; Pescaroli, G.; Graaff, H.; Felts, D.; Pietrantoni, L. Early warning and mass evacuation in coastal cities. *Coastal Eng.* **2014**, *87*, 193–204. [CrossRef]
27. Wang, Y. Advances in Remote Sensing of Flooding. *Water* **2015**, *7*, 6404–6410. [CrossRef]
28. Bodoque, J.M.; Amérigo, M.; Díez-Herrero, A.; García, J.A.; Cortés, B.; Ballesteros-Cánovas, J.A.; Olcina, J. Improvement of resilience of urban areas by integrating social perception in flash-flood risk management. *J. Hydrol.* **2016**. [CrossRef]
29. Birkholz, S.; Muro, M.; Jeffrey, P.; Smith, H.M. Rethinking the relationship between flood risk perception and flood management. *Sci. Total Environ.* **2014**, *478*, 12–20. [CrossRef]
30. Haer, T.; Botzen, W.J.W.; Aerts, J.C.J.H. The effectiveness of flood risk communication strategies and the influence of social networks—Insights from an agent-based model. *Environ. Sci. Policy* **2016**, *60*, 44–52. [CrossRef]
31. Opper, S.; Cinque, P.; Davies, B. Timeline modelling of flood evacuation operations. *Procedia Eng.* **2010**, *3*, 175–187. [CrossRef]
32. Jonkman, S.N.; Maaskant, B.; Boyd, E.; Levitan, M.L. Loss of life caused by the flooding of New Orleans after Hurricane Katrina: Analysis of the relationship between flood characteristics and mortality. *Risk Anal.* **2009**, *29*, 676–698. [CrossRef]
33. Ishigaki, T.; Onishi, Y.; Asai, Y.; Toda, K.; Shimada, H. Evacuation criteria during urban flooding in underground space. In Proceedings of the 11th International Conference on Urban Drainage, Scotland, UK, 31 August–5 September 2008; p. 7.
34. Matsuo, K.; Natainia, L.; Yamada, F. Flood and Evacuation Simulations for Urban Flooding. In Proceedings of the 5th International Conference on Flood Management, Tokyo, Japan, 27–29 September 2011; pp. 391–398.
35. Veeraswamy, A.; Galea, E.R.; Filippidis, L.; Lawrence, P.J.; Haasanen, S.; Gazzard, R.J.; Smith, T.E.L. The simulation of urban-scale evacuation scenarios with application to the Swinley forest fire. *Saf. Sci.* **2018**, *102*, 178–193. [CrossRef]
36. Zlateski, A.; Lucesoli, M.; Bernardini, G.; Ferreira, T.M. Integrating human behaviour and building vulnerability for the assessment and mitigation of seismic risk in historic centres: Proposal of a holistic human-centred simulation-based approach. *Int. J. Disaster Risk Reduction* **2020**, *43*, 101392. [CrossRef]
37. Ronchi, E.; Uriz, F.N.; Criel, X.; Reilly, P. Modelling large-scale evacuation of music festivals. *Case Stud. Fire Saf.* **2015**, *5*, 11–19. [CrossRef]
38. Kim, K.; Pant, P.; Yamashita, E. Integrating travel demand modeling and flood hazard risk analysis for evacuation and sheltering. *Int. J. Disaster Risk Reduction* **2018**, *31*, 1177–1186. [CrossRef]
39. Lämmel, G.; Grether, D.; Nagel, K. The representation and implementation of time-dependent inundation in large-scale microscopic evacuation simulations. *Transp. Res. Part C Emerg. Technol.* **2010**, *18*, 84–98. [CrossRef]
40. Bolia; Lakshay, N.B. Operating strategies of buses for mass evacuation. *Saf. Sci.* **2019**, *111*, 167–178. [CrossRef]
41. Lakoba, T.I.; Kaup, D.J.; Finkelstein, N.M. Modifications of the Helbing-Molnar-Farkas-Vicsek Social Force Model for Pedestrian Evolution. *Simulation* **2005**, *81*, 339–352. [CrossRef]
42. Lee, H.-K.; Hong, W.-H.; Lee, Y.-H. Experimental study on the influence of water depth on the evacuation speed of elderly people in flood conditions. *Int. J. Disaster Risk Reduction* **2019**, *39*, 101198. [CrossRef]
43. Bosina, E.; Weidmann, U. Estimating pedestrian speed using aggregated literature data. *Physica A: Stat. Mech. Its Appl.* **2017**, *468*, 1–29. [CrossRef]
44. Bernardini, G.; Quagliarini, E.; D'Orazio, M.; Brocchini, M. Towards the simulation of flood evacuation in urban scenarios: Experiments to estimate human motion speed in floodwaters. *Saf. Sci.* **2020**, *123*, 104563. [CrossRef]

45. Herman, I.P. Physics of the Human Body. In *Biological and Medical Physics, Biomedical Engineering*; Springer International Publishing: Cham, Switzerland, 2016; ISBN 978-3-319-23930-9.
46. Weidmann, U. *Transporttechnik der Fußgänger—Transporttechnische Eigenschaften des Fussgängerverkehrs (Literaturauswertung)*; Zweite ergänzte Auflage. Schriftenreihe des IVT 90, Institut für Verkehrsplanung, Transporttechnik, Strassen-und Eisenbahnbau, ETH Zürich: Zürich, Switzerland, 1993. [CrossRef]
47. Lu, X.; Yang, Z.; Cimellaro, G.P.; Xu, Z. Pedestrian evacuation simulation under the scenario with earthquake-induced falling debris. *Saf. Sci.* **2019**, *114*, 61–71. [CrossRef]
48. Parisi, D.R.; Dorso, C.O. Microscopic dynamics of pedestrian evacuation. *Phys. A Stat. Mech. Its Appl.* **2005**, *354*, 606–618. [CrossRef]

water

MDPI

Article

# Accounting for Uncertainty and Reconstruction of Flooding Patterns Based on Multi-Satellite Imagery and Support Vector Machine Technique: A Case Study of Can Tho City, Vietnam

**Sastry Dhara** [1,*], **Thanh Dang** [2], **Kajori Parial** [3] **and Xi Xi Lu** [1]

[1] Department of Geography, National University of Singapore, Singapore 119228, Singapore; geoluxx@nus.edu.sg

[2] Institute of Water and Environmental Research, Thuyloi University, Ho Chi Minh City 70000, Vietnam; dangducthanh@iwer.vn

[3] Geoinformatics & Remote Sensing Cell, West Bengal State Council of Science and Technology, Kolkata 700064, India; kajorigis@gmail.com

* Correspondence: sastry.dhara@u.nus.edu

Received: 31 March 2020; Accepted: 25 May 2020; Published: 28 May 2020

**Abstract:** One of the most frequent natural perils affecting the world today is flooding, and over the years, flooding has caused a large loss of life and damage to property. Remote sensing technology and satellite imagery derived data are useful in mapping the inundated area, which is useful for flood risk management. In the current paper, commonly used satellite imagery from the public domain for flood inundated extent capturing are studied considering Can Tho City as a study area. The differences in the flood inundated areas from different satellite sensors and the possible reasons are explored. An effective and relatively advanced method to address the uncertainties—inundated area capture from different remote sensing sensors—was implemented while establishing the inundated area pattern between the years 2000 and 2018. This solution involves the usage of a machine learning technique, Support Vector Machine Regression (SVR) which further helps in filling the gaps whenever there is lack of data from a single satellite data source. This useful method could be extended to establish the inundated area patterns over the years in data-sparse regions and in areas where access is difficult. Furthermore, the method is economical, as freely available data are used for the purpose.

**Keywords:** remote sensing; flood extent mapping; Can Tho City; Google Earth engine; uncertainty; support vector machine regression (SVR)

## 1. Introduction

Floods are one of the most frequent natural hazards affecting the world. Over the years, they have caused a large loss of life and damage to property [1,2]. Due to changing climates, it is observed that the magnitude and frequency of the floods are on the rise. In recent times, some of the events were severe enough to be termed as catastrophic [3]. Such events underlined the importance of estimating the flood risk and its management.

An important prerequisite for flood risk management is precise information on the real extent of inundation. There has been limited experience and recorded evidence available about the spatial extent of extreme floods. Traditional ground survey techniques have certain challenges, like a limited network of monitoring stations in some developing nations, inaccessibility of the area during floods, time consuming and laborious processes. [4,5]. In order to address these challenges of conventional surveys, remote sensing has emerged as an effective and efficient alternative technique. The ability of remote sensing data to provide a synoptic view of a large area has been found useful for identifying flood inundated areas [6]. It has become critical to use data derived from remote sensing technology along with hydrological models to better manage flooding risk [7].

With the increasing availability of satellite imagery data in the public domain, government agencies may use the remote sensing data for designing flood mitigation measures [8,9]. In the absence of reliable historical records, flood risk modelling helps us to look at possible extreme scenarios, while incorporating the changes in the flood risk profile with the help of remote sensing data [10–14]. Multilateral agencies such as the World Bank and the Asian Development Bank have used the results from flood risk assessments from time to time to explore various disaster risk transfer mechanisms [15,16].

The precise detection of dynamic changes in water levels and land cover change information to use for flood risk assessment depends on the spatial resolution of the image captured [17]. Various remote sensing products offer options in capturing the water boundaries over a large area at a certain point of time. Among them, high resolution satellite imagery data, such as Quickbird and WorldView images are available, but many of these data are not freely available [18]. Freely available optical imagery, such as Landsat Thematic Mapper (TM) and Enhanced Thematic Mapper (ETM+) can provide satellite imagery data at 30 m spatial resolution and an interval of 16 days. Archival records of satellite imagery data for more than three decades are available [19–21]. The longer-archived data and continual capture over the same areas makes the data very handy for mapping water bodies at a large regional scale for a variety of hydrological conditions. Band 7 data from Landsat is useful for distinguishing water bodies from surrounding land mass, with an error rate around 5%. There is another remote sensing satellite platform called Moderate Resolution Imaging Spectroradiometer (MODIS) (NASA, Washington, DC, USA) which includes two key instruments, Terra and Aqua satellites. Terra MODIS captures the imagery over the earth's surface every one to two days, and acquires the data in 36 spectral bands [8,9,15]. While Landsat data have relatively higher spatial resolution compared to data from MODIS, the revisit time of the Landsat sensor is longer than the MODIS. Remotely sensed images captured from the Sentinel-2 A and B optical sensors, launched by the European Space Agency in 2014, were also used in the current study. The revisit time of the Sentinel-2 sensor is about 10 days and the spatial resolution is between 10 to 60 m [10,16].

The Support Vector Machine (SVM) has been earlier used for applications such as land cover classification [22] in Can Tho City. The Google Earth Engine (GEE) became available later, which provides access to data from multiple satellite sensors via a single platform and also facilitates image processing functions which were earlier carried out in specialized remote sensing software (such as ERDAS/ENVI). The potential of GEE for analyzing flooded areas from the study area (Can Tho City) has not been studied much and this paper sheds light onto this important area. It is critical that such methods are explored in data scarce regions in SE Asia. Furthermore, some of these urban regions (such as Can Tho City) have experienced high economic growth in the recent years due to infrastructure development and migration from neighboring rural areas. There has been a scarcity of historical flood information, such as inundated area from ground surveys. In addition to establishing a procedure to retrieve the available historical information from GEE platform, our paper also proposes the use of SVM Regression, a relevant machine learning technique to address the spatial and temporal uncertainties in such regions. We believe that our paper proposes an important and economical solution which can be extended to establish historical flooding patterns for many SE Asian regions where such information is lacking.

Flooded water can be delineated using image processing techniques [11,12] on different satellite images, in the midst of increasing cloud cover. In order to gain the maximum benefit from satellite images, it is important to combine information from different kinds of sources, so that problems with one method can be overcome by employing the other methods [13,14]. However, advances in machine learning allow us to combine data from different satellites, thus providing an important option to minimize uncertainties in data resulting from different spatial and temporal resolutions. Here, we hypothesize that the use of an advanced machine learning technique of Support Vector Machine

Regression (SVR) can help identify the water-inundation areas over time when using satellite imagery with different spatio-temporal resolutions [23–25]. SVR also helps when incomplete data are available, thus enabling a more complete set of readings to be included into any further modelling. SVR can predict non-linear relationships and therefore used in the current context. The objective of this study is to minimize the uncertainties related to different spatial resolutions from different satellite platforms, by using SVR to reconstruct the flooding extents.

In this study, we used imagery from Landsat, MODIS and Sentinel-2 satellite sensors. These satellite platforms were chosen because of several reasons: 1) availability of data for long periods 2) data being available free of cost and 3) often two to three datasets would be available within the period from September to October (flooding season in Can Tho City) each year.

## 2. Materials and Methods

### 2.1. Study Area

The study area is Can Tho City, which is in the center of the Vietnamese Mekong Delta. Regions of the Vietnamese Mekong Delta are prone to high to extreme risk from floods [26–29]. The city is 75 km far from the Vietnamese East Sea (Figure 1). Can Tho City is one of the five cities centrally administered in Vietnam. It is the largest city in the Vietnamese Mekong Delta and an important base of increasing commerce, industry and transportation network [30,31]. As of April 2019, the city has a population of 1,235,171 [32] with a natural land area (in 2017) of about 1439 square km [33].

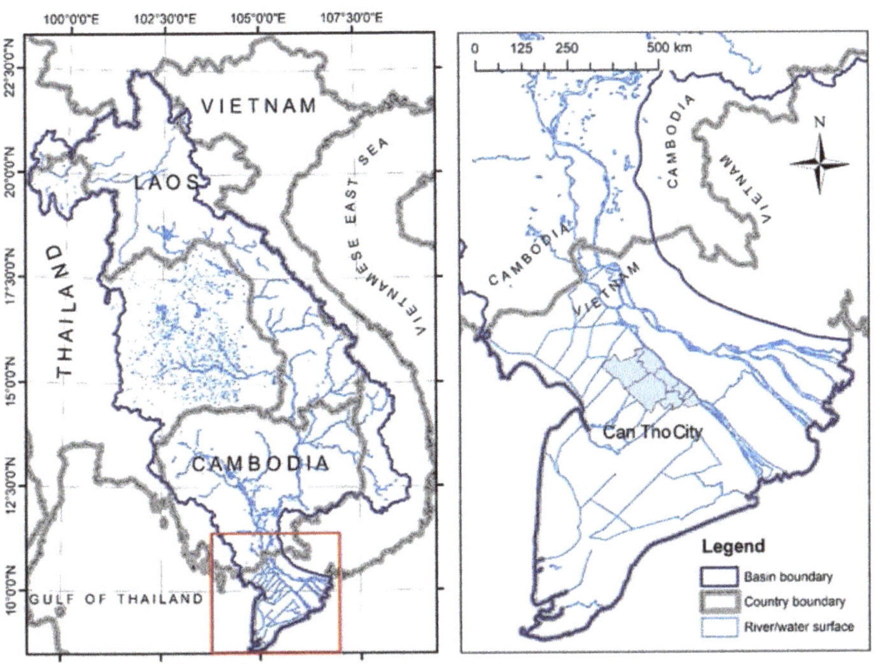


**Figure 1.** Lower Mekong Basin (**left figure**) in South-East Asia; Study Area Can Tho City (highlighted in the **right figure**).

The socioeconomic development goals for 2020 and 2030 of the Vietnamese Mekong Delta see the increasingly significant role of Can Tho City in achieving the growth objectives [29] Can Tho City is set to be the regional centre for many upcoming and developing industries, primarily based on the vast arable land and water bodies. These industries include aquaculture, fishing and food processing.

Figure 2 below shows the entrance to Tra Noc industrial zone in Can Tho City which houses several industrial firms. The increasing outputs from these industries in Can Tho City make the city a key contributor to the food security in the Mekong Delta region. Apart from these industries, Can Tho City plays an important role in the transportation connectivity via its many waterways, road network and the international airport. Due to the development of infrastructure as well as the industries, Can Tho City has attracted a lot of migrants in search of jobs and favorable economic engagement. As a growing city, it has witnessed a lot of construction activities in the recent times which resulted in changes in land use patterns. Figure 3 shows the ongoing construction activity of an industrial establishment. However, some of these economic growth related activities are hampered by the frequent natural disasters such as floods [34,35].

**Figure 2.** Tra Noc Industrial Zone in Can Tho City. The zone is home to industrial firms producing electronic, food processing, automotive, construction materials and pharmaceutical products.

**Figure 3.** An industrial building under construction in Can Tho City.

Analysis of flood risk in Can Tho City can be adapted to larger sections of South and South-East Asia since several of the geographical features with relation to proximity to major rivers and ocean as well as the weather patterns are noted in several developing cities in South and South-East Asia. Can Tho City faces threats from the three kinds of flooding: fluvial, pluvial, and tidal [36].

Upstream hydropower development along the Mekong may also cause hydrological alterations [37], but the impact on the delta is limited, with the change in flooding extents being slow to develop once they reach the lower parts of the delta. Importantly, the circumstances of the city are constantly changing and this brings about challenges and uncertainties in flood risk analysis and management.

Given the importance of Can Tho City to the economy, indeed to several aspects of the Vietnamese Mekong Delta, it is a good place to understand the development and maintenance of flood mitigation and management strategies in place. Can Tho City faces multiple challenges such as the increase in sea level and river runoff due to climate change, urban runoff caused by imperviousness and potential intensification of extreme rainfall due to microclimatic shifts [38]. Floods in Can Tho City cause losses to the city's infrastructure and result in an adverse impact to the livelihoods of citizens. Some of the reasons for the flooding in Can Tho City include the lack of robust flood-prevention system and rapid urbanization [39].

Can Tho City is also an ideal place to study the effect of uncertainties in current flood risk assessment, which can eventually lead to strategies for tackling these uncertainties. In time, strategies developed in an accessible—but developing—city, can be adapted to larger sections of South and South East Asia.

Figure 4a–d below capture some of the scenes observed in the streets of Can Tho City during the floods in the year 2014. Due to the varied processes via which flooding can occur in the Mekong Delta, the Vietnamese government is attempting to provide plans to mitigate damage from flooding [29]. One way to do this would be to assess inundation patterns from the past. There is a paucity of on-the-ground data regarding flood extents in these areas. This has led to greater reliance on data gathered from remote sensed data. Remotely sensed images from satellites are used to reconstruct historical flood extent maps in several nations [40,41]. With the different kinds of satellites in use for accessing these data, reliability and usability of such data should be studied. One of the ways would be via examining the potential of freely available satellite imagery for the study area. Here, we used the capacity of the Google Earth Engine (GEE), a web-based platform, to analyze the flooded areas from three different satellite imagery (Landsat, MODIS and Sentinel-2). Each of these three satellite image sensors has their own advantages and disadvantages (e.g., different spatial and temporal resolutions). A range of potential flooded areas are provided by each of these sensors, and SVM was used to reconstruct the flooding patterns.

**Figure 4.** (**a–d**) Photographs showing the different inundated areas in Can Tho City during the floods in the year 2014 (Courtesy: Dr Dunja Krause, United Nations Research Institute for Social Development (UNRISD, Geneva, Switzerland).

### 2.2. Methodology

The below steps were followed to conduct the study. Further description about the data and techniques are provided in Sections 2.2.1–2.2.3, 2.3 and 2.4.

a. Available satellite imagery from Landsat and Terra MODIS sensors covering the study area between the years 2000 and 2018 were accessed via the GEE platform. Available imagery from Sentinel-2 sensor is accessed from the year 2015 onwards.
b. Extent of water bodies from respective satellite imagery are calculated using Normalized Difference Water Index (NDWI). Details are outlined in Section 2.2.2.
c. The calculated inundated area figures (i.e., extent of water bodies) from different satellite sensors provide different values. This is due to the spatial and temporal differences of the three different satellite sensors which were considered. To account for this uncertainty, Support Vector Machine Regression (SVR) technique has been used to reconstruct the inundated area values for different months as an integrated value from three sensors.

#### 2.2.1. Google Earth Engine

Google Earth Engine (GEE) is a web-based cloud computing platform capable of storing multi-layer catalog of satellite images and geospatial datasets within GEE's Public Data Catalog [42–44]. In addition to archival, within the GEE script-environment these large datasets can be analyzed for identifying the changes, mapping variations, and variances on the surface of the earth. Since GEE is a cloud-based service, there is no necessity to download and analyze imagery in the traditional manner and thus this saves time.

Phongsapan and colleagues studied the potential of GEE platform in deriving an operational flood risk index covering Myanmar [45]. Uddin et al. outlined the utility of GEE in using multi-temporal Sentinel-1 Synthetic Aperture Radar (SAR) satellite imagery in capturing the flood inundated areas in Bangladesh [46]. Sidhu and team and Celik and team also used GEE for identifying the land cover changes in urban areas in Singapore [47] and Ankara [48], respectively.

Nguyen and colleagues analyzed several Landsat satellite imageries to examine the changes in built-up area in Can Tho City between the years 1998 and 2018. However, GEE was not used to retrieve and analyze the Landsat imagery in this case [49]. Goldblatt et al. earlier demonstrated the application of GEE for identification of boundaries of urban regions (i.e., built-up areas) in India [50]. Review of studies such as the above illustrate that there is a gap in research and utilization of GEE for research work in Can Tho City (study area), more so in the important area of flooding risk. The important idea of utilization of historical satellite imagery via GEE, an efficient archival and computational platform in Can Tho City was introduced in our paper.

Researchers wishing to use the service neither need to be familiar with nor need to use specialized remote sensing software such as Environment for Visualizing Images (ENVI) and Earth Resources Data Analysis System (ERDAS) Imagine for regular image processing functions. As with other open source software, several algorithms are provided by other users, thus providing a greater expertise in analyzing different datasets than would be possible for a single person or team working together. Bulk downloads and memory use can be avoided since work can be performed online. In this study, datasets related to Can Tho City from three publicly available platforms, Landsat (5, 7 and 8), Sentinel-2 and MODIS were obtained and analyzed using the Google Earth Engine.

#### 2.2.2. Extraction of Water Body Extent

The Normalized Difference Water Index (NDWI) [42,51,52] is an index derived from the satellite imagery using the Near-Infrared (NIR) and Green wavelengths. NDWI can capture the water body presence from remote sensing imagery via separating non-water related features. Reflected near-infrared radiation and green light are used in the calculation since this helps in excluding vegetation and land, while improving water detection [52]. NDWI is calculated with the formula as shown below:

$$\mathrm{NDWI} = \frac{\mathrm{Green} - \mathrm{NIR}}{\mathrm{Green} + \mathrm{NIR}} \tag{1}$$

where the Green is the band which captures reflected green light and the NIR represents the near-infrared radiation.

By using green wavelengths, the typical reflectance of water features is maximized, the low reflectance of NIR by water bodies is minimized and the high reflectance of NIR by vegetation and soil features from the land is maximized. The outcomes from the above index are that water bodies have positive values while using multispectral imagery which has a reflected green band and an NIR band. Soil and vegetation features have zero or negative values due to the typical high reflectance of NIR than the green light [52].

During the processing of satellite imagery, an appropriate threshold was applied based on the presence and extent of cloud cover in the imagery. For example, a threshold value of 0.1 was applied for imagery with high cloud coverage whereas a threshold value of 0.3 was applied for the satellite imagery with relatively less cloud coverage [53]. This helps in capturing even very low values of NDWI which may result from high cloud coverage or the absence of water bearing pixels. When the water bodies are clearly visible, the increase in threshold value beyond 0.3 does not have much effect; hence a threshold value of 0.3 is good enough for imagery with clear water body presence and less cloud cover.

#### 2.2.3. Data (Landsat 5,7,8, Terra MODIS and Sentinel-2)

Landsat

The Landsat programme is a joint venture of United States Geological Survey (USGS), National Aeronautics and Space Administration (NASA) and National Oceanic and Atmospheric Administration (NOAA), and has been in use since 1972 [20,21]. For most Landsat imagery, the temporal resolution (sensor revisit time) of Landsat is 16 days. For Landsat imagery that includes multi-spectral and thermal data, spatial resolution is mostly about 30 m. Landsat data are available in the GEE in its raw form, surface reflection, Top Of Atmosphere (TOA)-corrected reflection, and other ready-made products, such as NDWI, NDVI and EVI indices. Landsat sensor, however, cannot capture the data when there is a cloud cover. From Figure 5a below, it can be noted that parts of the study area, covered with clouds are captured as white patches. This posed issue with derivation of accurate inundated area during certain periods. Scanline collector failure of the Landsat 7 Enhanced Thematic Mapper in 2003 led to data gaps with 22 percent of the coverable area missing. Figure 5b below illustrates this issue. Images acquired in these two cases (with cloud cover and scan line cover) introduce some uncertainty in the inundation area captured. Figure 5c,d below show the false color Landsat images acquired during dry and flood season respectively.

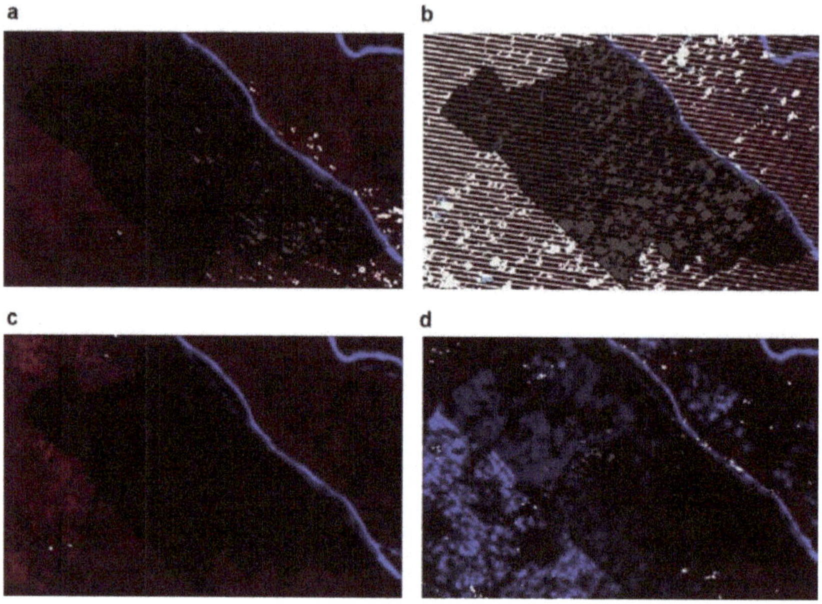


**Figure 5.** False color Landsat imagery showing the study area during (**a**) cloud cover (**b**) scan line collector failure (**c**) dry season (**d**) flood season.

MODIS

The Moderate Resolution Imaging Spectroradiometer (MODIS) was built by Santa Barbara Remote Sensing facility and was launched into the orbit by NASA during the year 1999 on board the Terra (EOS AM) Satellite, and in 2002 on board the Aqua (EOS PM) satellite [15,54]. These instruments are capable of imaging the surface of the Earth every 1 to 2 days, at 250 m, 500 m and 1 km resolutions. Data are captured in more than 35 spectral bands, ranging from <1 μm to nearly 15 μm. Changes and processes occurring in the oceans and on land can be captured by MODIS. Figure 6a,b show the False color Terra MODIS images acquired during dry and flood seasons respectively.

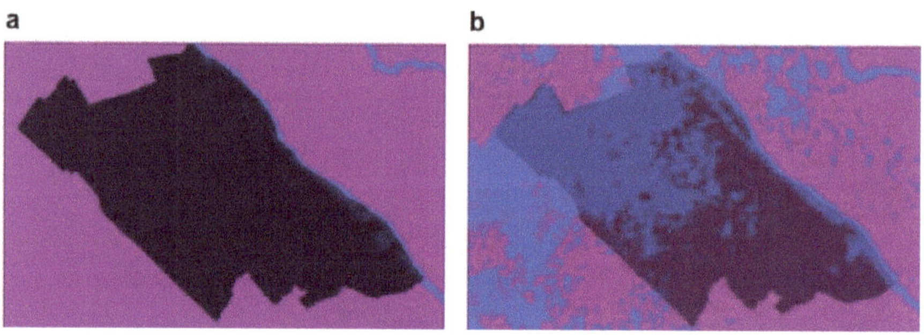


**Figure 6.** False color Terra Moderate Resolution Imaging Spectroradiometer (MODIS) image showing the study area during (**a**) dry season and (**b**) flood season.

Sentinel-2

Sentinel is a group of satellites managed by the European Space Agency (ESA, Paris, France) under the Copernicus program in 2014 [10,16,55]. Here, we used high-resolution optical images from

Sentinel-2A and 2B, which together have a temporal resolution of about 5 days and a spatial resolution of 30 to 60 m. Figure 7a,b show the false color Sentinel images showing the study area captured during dry and flood season respectively.

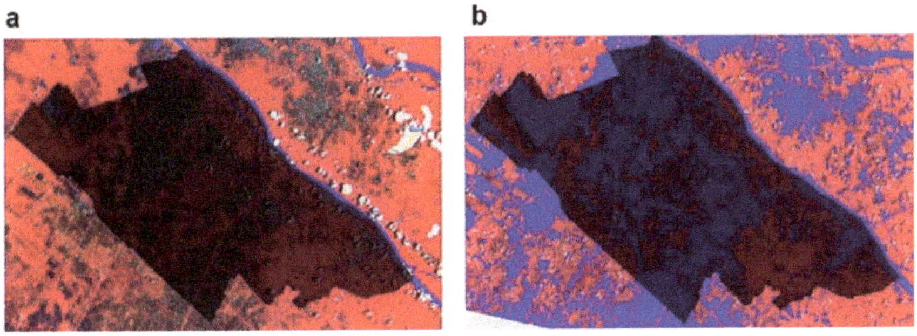

**Figure 7.** False color Sentinel-2 image showing the study area during (**a**) dry season and (**b**) flood season.

## 2.3. Data Time-Period

In this study, data from the years 2000 to 2018 were collected for the study area from Landsat (5,7,8) and Terra MODIS. We attempted to collect one data point from the data from each month in this time period (total of 228 possible months). Certain prerequisites needed to be maintained when remotely sensed images were collected for analyses of flooding extent, like less than 20% cloud coverage (0% if possible, since floods occur during the rainy season) of the images. Landsat provided spotty images between 2003 and 2014, due to the partial failure (scan line collector failure) of the Landsat 7 satellite [56,57]. Additionally, the Sentinel-2 satellite was launched recently, and data are only available from the year 2015 onwards.

Here, we used the classification of pixels in the remote sensing image into those containing water and those not containing water by using the NDWI. This was undertaken for all the available and usable images from Landsat, MODIS and Sentinel-2 sensors. Of the possible 228 data points from each type of sensor, data from 82 months is available from Landsat and data from 214 months is available from MODIS. For Sentinel, data are only available from September 2015, and of the possible data from 40 months, we obtained data available from 27 months.

While analyzing data from different platforms, discrepancies are noted in the inundated areas obtained from each platform. We therefore used support vector regression (SVR), a non-linear regression technique, to reconstruct flooding patterns.

## 2.4. Support Vector Machine Regression (SVR)

SVM has earlier been used mainly as a supervised classification technique while analyzing satellite imagery [58]. It is a supervised, non-parametric classification algorithm, used for linear binary classification of data points [23,59]. Support vector refers to data points that fall along the border of the margin of separation. As with any supervised classification algorithm, SVM requires a training data set. However, with SVM, a small training set is enough and can provide good classification accuracy. For every data point, a decision is made whether it is far from the support vector. The distance from the support vector determines the margin of classification. It classifies each new datapoint from the testing set without making assumptions as to the classification, i.e., data classification is not dependent on assumption of Gaussian distribution and not based on nearest neighbors. SVMs also provide a balance between data overfitting and underfitting, even when limited training samples are used. This method is often used to for images obtained from multi-spectral satellite sensor images.

The potential of Support Vector Machine (SVM) as a supervised machine learning classifier in deriving the flooded area from Landsat imagery has been studied by Ireland and colleagues [60].

Syifa et al. utilized SVM and Artificial Neural Network (ANN) classification techniques to derive the map showing the flooded area after the collapse of Brumadinho dam wall in Brazil during January 2019 [61]. However, the SVM technique can be applied for both classification and regression. In the current study, we applied the principles of SVM for regression, called Support Vector Machine Regression or simply Support Vector Regression (SVR). Gizaw and Gan applied SVR technique for the estimation of regional flood frequency covering two study areas in Canada [62]. Chen and Yu adopted SVR to develop a model to provide a deterministic flooding stage forecasting for Lang-Yang river in Taiwan [63]. SVM and SVR have become effective machine learning substitutes to Artificial Neural Networks (ANNs), finding more applications in flood prediction [64]. These methods have been applied in several flood prediction studies with better performance results compared to ANN and Multiple Linear Regression (MLR) techniques. These methods were applied involving data such as flood time series, extreme rainfall and streamflow [64,65].

For the training set, data from homogenous areas in the collected images were utilized. Here, SVR was used to adjust for the missing data from several months. When data from at least one sensor are available, SVR can generate a composite value for the inundated area. The major reason for using SVR is the different spatial and temporal resolutions for the three satellite imagery we analyzed here. While Landsat and Sentinel-2 imagery provide relatively high spatial resolution, they have lower temporal resolution. In addition, SVR can better predict non-linear predictor/predictand data as used here, and which cannot be well predicted via linear regression methods.

In total, we analyzed 83 Landsat images, 214 MODIS images and 27 Sentinel images. Where available, for each Landsat or Sentinel image, we found a corresponding MODIS image and there were 105 image pairs. We bootstrapped these pairs with the training and the testing ratio of 7:3, 100 times. The average value of coefficient of determination ($R^2$) for 100 rounds of this was 0.85 for training, and 0.72 for prediction. By using this method, we were able to reconstruct integrated flood extent values from MODIS data using the SVR model constructed from the image pairs. Packages e1071 and boot were used for these analyses, performed on R [66–69].

## 3. Results

### 3.1. Capturing Inundated Areas Using GEE

In this study, "ground-truth" information from the field survey was not available for the study area under consideration. This issue is not uncommon for flooding risk and this often results in scarcity of quantitative validation [70]. To address this challenge, we collected 200 random high-resolution Google Earth images (100 water and 100 non-water images). The Kappa Coefficient of Landsat, MODIS and Sentinel-2 accessed from GEE were 0.87, 0.75 and 0.89 respectively. These Kappa index values demonstrate strong (Kappa value > 0.8) to moderate (Kappa value > 0.7) correlation between inundated areas as determined by satellite imagery from GEE and an external source, thus allowing for the use of satellite images and the corresponding thresholds in classifying the inundated areas/water bodies. The lower K value of MODIS compared to the two others may be caused by its lower spatial resolution.

### 3.2. Correlation and Differences Between Data from Different Satellite Sources

In this study, data from different satellites were utilized to derive the extent of water inundation in Can Tho City. As an initial step, it is necessary to assess the degree of agreement between the flood extents derived from the different satellite images. As such, we performed Pearson's correlation between data from Landsat and MODIS images. The data are highly correlated, with a Pearson's r of 0.84 (Figure 8a). We also performed Pearson's correlation between Landsat and MODIS data separately for the low inundation period (from the months of February to July) and the high inundation period (August to January). For this process, we collated the Landsat data from the months of February to July throughout the study period and correlated it with the MODIS data collected from February to July during the study period. A similar correlation was performed on data collected from August to January. The results showed that the Pearson's correlation coefficient was low during the low inundation period (r = 0.15, low inundation period). However, the correlation was high during the high inundation period (r = 0.79, high inundation).

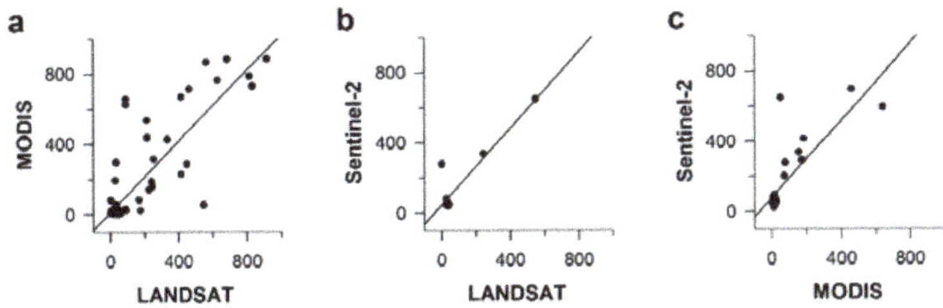


**Figure 8.** Correlation of inundation data between different sensors. Monthly inundated area captured from one sensor is plotted on the $x$-axis and the other is plotted along the $y$-axis. (**a**) Left picture: Landsat and MODIS Sensors; (**b**) Middle picture: Landsat and Sentinel-2 Sensors and (**c**) Right picture: MODIS and Sentinel-2 Sensors.

Similarly, Pearson's correlation analysis was performed between data from Landsat and Sentinel-2 images (Figure 8b) and MODIS and Sentinel-2 images (Figure 8c). These data are also highly correlated, with a Pearson's r of 0.91 and 0.80, respectively. Limited data are available from Sentinel-2, since it was launched more recently than the others. Its correlation with the Landsat data is very high, given the similarly high spatial resolutions of both these data.

A major reason for the differences between the data collected from the three sensors is the differences in spatial resolution of the data collected. Additionally, cloud coverage, especially during the times of high inundation, is another important factor. High inundation happens during the rainy season, and therefore completely avoiding cloud cover would be nearly impossible. Nevertheless, the great degree of correlation informs us that over a long period of time, data from different sensors can similarly inform us about the water-inundated area over several years.

Unfortunately, the "ground-truth" information from the field survey is not available for the study area under consideration. This issue is not uncommon for flooding risk and this often results in scarcity of quantitative validation [70].

### 3.3. Seasonal Variations in Water Levels Observed by Different Sensors

As stated in the previous section, the correlation between Landsat and MODIS data was lower during the months of low inundation. To assess the agreeability between the different sensor data, we also plotted the time course of the Landsat, MODIS and Sentinel-2 data over the full time period of analysis, from 2000 to 2018. We observed that the data collected from the different sensors show some differences. However, the results from all the sensors demonstrate increase in the inundation area during the months of increase in rainfall and decrease in months when there is decrease in rainfall (Figure 9).

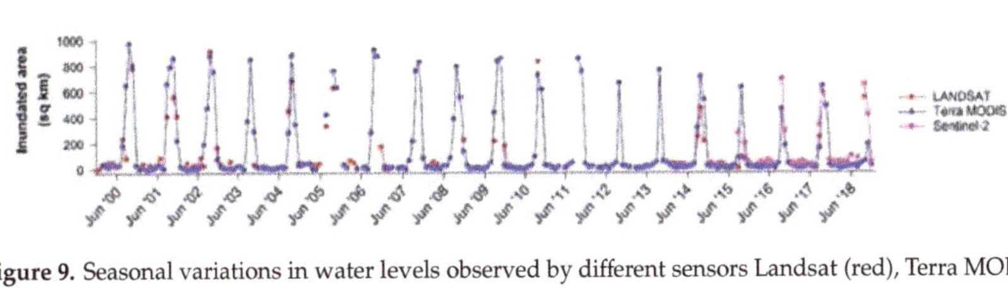


**Figure 9.** Seasonal variations in water levels observed by different sensors Landsat (red), Terra MODIS (blue) and Sentinel-2 (pink).

### 3.4. Use of Support Vector Machine Regression (SVR) for Reconstructing Inundation Areas

The correspondence of inundation areas derived from Landsat and Sentinel-2 imagery is observed to be high (Figure 8b), as can be expected based on the high correlation (Pearson's r = 0.91). Hence, we used the Landsat and Sentinel-2 data to train the MODIS data using the SVR approach.

Reconstructed values were obtained for inundation during all the months when images from MODIS were available. Inundated area values obtained from the sensors and reconstructed data are plotted in Figure 10 below. The MODIS data from the 105 image pairs are trained from the corresponding Landsat/Sentinel-2 data (Figure 10a, red dots) via SVM regression and the reconstructed data derived (Figure 10a, blue dots). There is a reduction in the variation in the reconstructed values compared to those from MODIS data (Figure 10b).

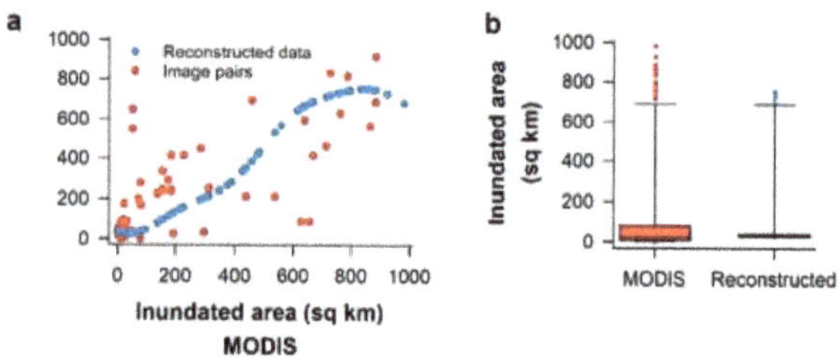


**Figure 10.** Reconstructed data from Support Vector Machine Regression (SVR) analysis (**a**). Comparison of image pairs used for Support Vector Machine (SVM) regression analysis and reconstructed data. The 105 image pairs used to train the SVM regression model are plotted in red and the SVR results are plotted in blue. The $x$-axis indicates original inundated area in sq km derived from Terra MODIS and $y$-axis indicates inundated area data in sq km from Landsat/Sentinel-2 (for the red dots) and from reconstructed data (for the blue dots); (**b**). Box plots showing the inundated area in square kilometers from Terra MODIS sensor (left) and the reconstructed value (right). The $y$-axis shows inundated area (MODIS data and reconstructed values) in sq km.

When the reconstructed data are plotted against time, the pattern of the data shows increases in the water-inundated area during the wet months and decreases during the dry months, when water flows only in the waterways, as is seen with the original data (Figure 11). This training approach ensures that data from more than one sensor can be used to generate a composite model of the inundation areas during all the months when data from at least one sensor is available.

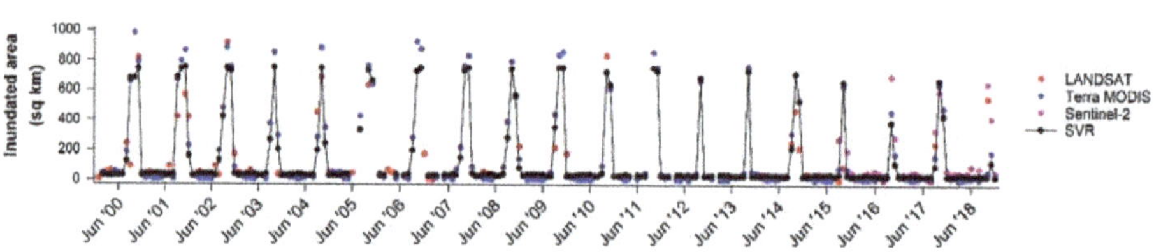


**Figure 11.** Reconstructed area of flood inundation using support vector machine regression (black).

### 3.5. Trend of Inundation Areas in Can Tho City

During the years of the study, it was noted that the total inundation area has steadily decreased in Can Tho City. To better observe this, we obtained the average monthly inundation every year using the reconstructed SVR data (Figure 10). In the graph shown below (Figure 12), the average monthly inundated area (in square kilometers) reconstructed using Support Vector Machine technique is plotted on the $y$-axis against the respective years from 2000 to 2018 on the $X$-axis. The seasonal Mann-Kendall test was used to test this. The Mann-Kendall Tau was −0.1003, with a $p$ value of 0.03.

This indicates that the inundation decreased overall during the time period of the study. After the year 2000, the Vietnamese government invested more in the dyke system to increase agricultural production (rice) in Vietnamese Mekong Delta (VMD). This resulted in a general decrease in flooding area over the years. A similar pattern has been reported in studies conducted in the other parts of VMD, such as the Long Xuyen Quadrangle (LXQ), the Plains of Reeds (POR) and An Giang province of the Vietnamese Mekong delta [34]. Though it is not a direct validation, the similarity in the trends with the other regions provides the consistency in the effects after the investment in dyke systems from the year 2000.

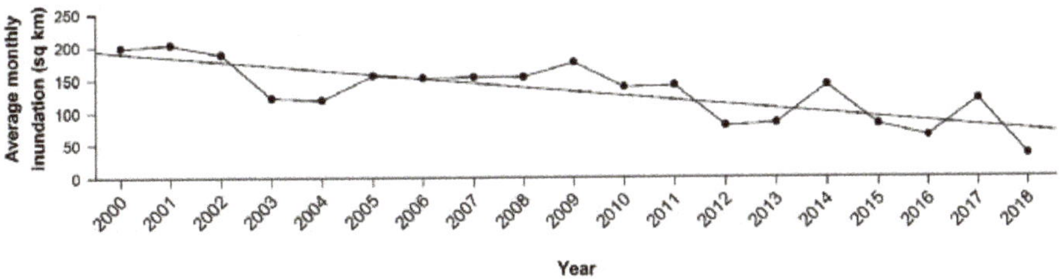


**Figure 12.** The reconstructed average monthly inundation area in Can Tho City from 2000 to 2018. Inundation is decreasing over time similar to that observed in other parts of the Vietnamese Mekong Delta [34].

## 4. Discussion

In this study, we analyzed images of Can Tho city from three different satellites (Landsat, MODIS and Sentinel-2) with varying temporal and spatial resolutions, and used a non-linear regression method—Support Vector Machine Regression—to integrate all the data an derive one single value. Here, we show that the correlation between data from different satellites is high, allowing for the training of the SVM to obtain inundation data for the period of 19 years, from 2000 to 2018. We also observe that the inundation area has been steadily decreasing over the years. Our results provide a proof-of-concept that machine learning algorithms like SVM can be successfully applied to combine data from different satellites to obtain one seamless inundation result that can be used for flood risk management. The results show that this can be completed with the use of freely available data and cloud computing platforms, thus allowing economically viable solutions for rapidly developing regions in South East Asia, such as those exemplified by Can Tho City.

Several studies have been carried out on flood risk assessment, via the use of satellite imagery. However, use of satellite data leads to the introduction of uncertainties. This is because of the different spatial and temporal resolutions of the satellite data. In addition, the failure of Landsat scan line corrector from May 2003 adds another layer of uncertainty, leading to lack of data for a certain period. These uncertainties are the epistemic uncertainties seen in flood risk assessment, that are introduced due to a lack of sufficient data and knowledge [71]. These uncertainties can be reduced by combining data from different satellites as we did in this study, and by using the power of machine learning to resolve these issues, which is a step in the right direction. While it is tempting to compare the data from our current study to historical flood extent data, this was not carried out, because the historical flood extent data are obtained from Landsat data, and thus cannot be compared here.

Here, we also showed that GEE can be used successfully to capture the inundated areas. The cloud computing capacity of GEE can be used to analyze massive amounts of data in a relatively short period of time. One other advantage is that the GEE platform provides data from different satellites under one platform. While GEE has mostly been used to detect changes in land cover, here, we used the engine to analyze data related to inundated regions. From the reconstructed average monthly inundation between the years 2000 and 2018, it is noted that there is a general decreasing trend. The possible reasons for this decreasing trend include urban expansion (i.e., urban built up area) and increase in the local flood protection systems [22,72]. Furthermore, more water is transferred to the middle of the

Vietnamese Mekong Delta due to the upstream flood prevention systems and sea level rise. [27,73,74]. There is some uncertainty in the inundated area capture due to the cloud cover in some satellite imagery and the operation of sluice gates. However, the effect of this uncertainty on the long-term inundation pattern is minimized by using the data from three different satellites (Landsat, MODIS and Sentinel-2) and SVM regression approach.

The temporal resolution of MODIS satellite sensor is around two days, whereas the satellite revisit periods of Landsat and Sentinel-2 are 16 and five days respectively. Furthermore, Sentinel-2 data only became available from September 2015. Landsat sensor provided spotty images between 2003 and 2014. Data from Terra MODIS are therefore more frequently available than the other two sensors, although its spatial resolution is relatively low. Hence, the data were available for certain months from the three different sensors, but there are cases where the inundated area data are available only from MODIS. We used the SVM technique to integrate the three techniques to output a single inundation value, a reconstructed inundated area figure when two or three different figures are available.

Satellite remote sensing imagery derived data have been increasingly recognized as a critical source for identifying land features and changes in the land cover from time to time. There are several freely available satellite data sources and the user community, particularly engaged in flood risk management, could benefit from this open source data. Whilst there are differences in the temporal and spatial resolutions from different data sources, appropriate techniques could be applied to derive optimal advantage from this disparate sets of data. For example, advanced data science techniques like machine learning could be employed to "fill-in" the data gaps while using the available records from multiple sources. This information is quite useful in the sense that it helps provide an indicative reference information based on scientific methods. During floods, it is not uncommon that a robust mechanism is not always available or possible to capture the "ground-truth" or actual flood extent map [75]. This issue is prevalent in many Asian territories, including Vietnam. The reasons for this include the difficulties in logistics of field surveying, and issues with reaching inaccessible areas during or immediately after a flooding event [5].

In the absence of ground-truthing data, alternative methods such as the usage of crowdsourcing data (photographs), development and deployment of mobile applications for automatic inundated area detections and analyzing the information shared via social media platforms (e.g., geotagged images) can be considered [76]. The role of people living in the flooded areas is crucial for information dissemination about the flood event. They can help in verifying the data collected and contribute to the reports when flood detection sensors are not available.

The contribution from citizens and development of integrated platforms to collect and analyse the data can help in improving the granularity of the data, such as both the temporal and spatial resolutions, and can be an effective form of the validation of data from other sources like remote sensing satellites. Some good examples of such initiatives include WeSenseIt, Ground Truth 2.0, and FloodCitiSense projects funded by European Union [77]. The accretion of crowdsourcing data for understanding flooding events is vital, even more so when the events occur in urban regions. The limitations in the temporal resolution from remote sensing sensors can be overcome as the crowdsourcing information can be made immediately accessible. Furthermore, flood depth information can also be collected at various places of interest at regular intervals to augment the data from other sources [78].

The recent improvements in remote sensing and geographical information systems (GIS) technologies have paved the way for further innovations in natural disaster risk management, particularly in flood risk assessment, damage assessment and planning [79]. There could be newer machine learning methods or improvisations to potentially enhance the SVR going forward. Further studies in this direction are important.

## 5. Conclusions

Flooding in Can Tho City can cause huge economic losses and impact many lives. In this study, a cost-effective solution for the study of flood risk via the usage of freely available satellite imagery

(Landsat, MODIS and Sentinel-2) was proposed. Google Earth Engine, a powerful satellite imagery data archival and computing platform, was used to derive the areal extents of the inundated area for every month over a period between the years 2000 and 2018. Support Vector Regression (SVR), a supervised machine learning algorithm has been used to reconstruct the water-inundated areas. This reconstruction allowed for the filling of the data gaps while considering the data uncertainties from different satellite sensors with various temporal and spatial resolutions. This also helped establish a relationship for identifying the flooding pattern over the years. The inundated area pattern allows the users to examine the trends in flooding and assess the flood risk. This is a useful technique which could be extended to several other SE Asian cities and provinces which are prone to flooding.

Remote sensing can provide spatially continuous data, unlike point measurements, from gauging stations, and this is especially useful in the case of long duration floods [80]. Satellite revisit time allows for flood monitoring at regular intervals without a physical presence at the location of flooding, especially when access is difficult. In addition, the remote sensing data are easy to access, and are rapidly processed. However, these data carry some limitations; the quality and applicability is subject to the spatial resolution and satellite repeat visit time. There is also scope for errors in identifying the flooded area based on the image processing algorithm used. If the revisit time of the satellite is longer, then the monitoring is not possible at short intervals of time [14]. Nevertheless, the advantages of these data far outweigh the disadvantages and with improved techniques and sensors, these disadvantages can be overcome.

In this study, we provided a proof-of-concept, showing that using machine learning algorithms can help integrate data from three different satellite sensors to derive one single integrated value, providing economically more feasible methods in regions of the world undergoing rapid growth. A small number of data pairs can be sufficient to do so, especially when techniques such as bootstrapping are utilized along with support vectors, as was carried out here. Since a greater number of MODIS images are available due to its relatively higher temporal resolution, we propose that Landsat and/or Sentinel-2 data can be used to train MODIS to derive the composite inundation values. In this way, the advantages of data from greater temporal and spatial resolution may be integrated by the researchers and other practitioners. Here, all the data and analytical tools we used are freely available. We hope that others can use similar methods to access data in places where field data are not easily available, but where the assessment of flood risk is critical for planning and development purposes.

**Author Contributions:** Conceptualization, S.D.; Data curation, S.D.; Formal analysis, S.D. and T.D.; Investigation, S.D.; Methodology, S.D. and T.D.; Project administration, S.D.; Resources, S.D.; Software, S.D. and T.D.; Supervision, X.X.L.; Validation, S.D.; Visualization, S.D., T.D. and K.P.; Writing—original draft, S.D.; Writing—review & editing, S.D., T. D. and K.P. All authors have read and agreed to the published version of the manuscript.

**Funding:** This research received no external funding.

**Acknowledgments:** We acknowledge the support from Van Pham Dang Tri, Can Tho University and Pham Thi Mai Thy, Vietnam National Space Centre for their support in connecting with the relevant research personnel and sharing the local knowledge. Thanks to Dunja Krause, United Nations Research Institute for Social Development (UNRISD) for allowing to use her pictures. We appreciate the support from Nguyen Hanh Quyen and her team at Asian Disaster Preparedness Centre, Thailand for sharing useful inputs related to Google Earth Engine. We finally thank the Google Earth Engine development team and community for their support through their forum.

**Conflicts of Interest:** The authors declare no conflict of interest.

## References

1. Alfieri, L.; Burek, P.; Feyen, L.; Forzieri, G. Global warming increases the frequency of river floods in Europe. *Hydrol. Earth Syst. Sci.* **2015**, *19*, 2247–2260. [CrossRef]
2. Schiermeier, Q. Increased flood risk linked to global warming. *Nature* **2011**, *470*, 316. [CrossRef] [PubMed]
3. Akstinas, V.; Meilutytė-Lukauskienė, D.; Kriaučiūnienė, J.; Šarauskienė, D. Features and causes of catastrophic floods in the Nemunas River basin. *Hydrol. Res.* **2019**, *51*, 308–321. [CrossRef]
4. Sanyal, J.; Lu, X.X. Application of Remote Sensing in Flood Management with Special Reference to Monsoon Asia: A Review. *Nat. Hazards* **2004**, *33*, 19. [CrossRef]

5. Rahman, M.R.; Thakur, P.K. Detecting, mapping and analysing of flood water propagation using synthetic aperture radar (SAR) satellite data and GIS: A case study from the Kendrapara District of Orissa State of India. *Egypt. J. Remote Sens. Space Sci.* **2018**, *21*, S37–S41. [CrossRef]
6. Refice, A.; D'Addabbo, A.; Lovergine, F.P.; Tijani, K.; Morea, A.; Nutricato, R.; Bovenga, F.; Nitti, D.O. Monitoring Flood Extent and Area Through Multisensor, Multi-temporal Remote Sensing: The Strymonas (Greece) River Flood. In *Flood Monitoring through Remote Sensing*; Refice, A., D'Addabbo, A., Capolongo, D., Eds.; Springer: Berlin/Heidelberg, Germany, 2018. [CrossRef]
7. Wang, X.; Xie, H. A Review on Applications of Remote Sensing and Geographic Information Systems (GIS) in Water Resources and Flood Risk Management. *Water* **2018**, *10*, 608. [CrossRef]
8. Revilla-Romero, B.; Hirpa, F.; Pozo, J.; Salamon, P.; Brakenridge, R.; Pappenberger, F.; De Groeve, T. On the Use of Global Flood Forecasts and Satellite-Derived Inundation Maps for Flood Monitoring in Data-Sparse Regions. *Remote Sens.* **2015**, *7*, 15702–15728. [CrossRef]
9. Arvind, C.S.; Vanjare, A.; Omkar, S.N.; Senthilnath, J.; Mani, V.; Diwakar, P.G. Flood Assessment using Multi-temporal Modis Satellite Images. *Procedia Comput. Sci.* **2016**, *89*, 575–586. [CrossRef]
10. Rapinel, S.; Mony, C.; Lecoq, L.; Clément, B.; Thomas, A.; Hubert-Moy, L. Evaluation of Sentinel-2 time-series for mapping floodplain grassland plant communities. *Remote Sens. Environ.* **2019**, *223*, 115–129. [CrossRef]
11. Mohd, M.I.S.; Mansor, M.A. Application of Remote Sensing and Hydrological Modelling in Flood Prediction Studies. *Malays. J. Remote Sens. GIS* **2000**, *1*, 91–98.
12. Samarasinghe, S.M.J.S.; Nandalal, H.; Weliwitiya, D.P.; Fowze, J.S.; Hazarika, M.; Samarakoon, L. Application of Remote Sensing and GIS for flood risk analysis: A case study at Kalu-Ganga River, Sri Lanka. *Int. Arch. Photogramm. Remote Sens. Spatial Inf. Sci.* **2010**, *38*, 110–115.
13. Giustarini, L.; Chini, M.; Hostache, R.; Pappenberger, F.; Matgen, P. Flood Hazard Mapping Combining Hydrodynamic Modeling and Multi Annual Remote Sensing data. *Remote Sens.* **2015**, *7*, 14200–14226. [CrossRef]
14. Lin, L.; Di, L.; Yu, E.G.; Kang, L.; Shrestha, R.; Rahman, M.S.; Tang, J.; Deng, M.; Sun, Z.; Zhang, C.; et al. A review of remote sensing in flood assessment. In Proceedings of the 2016 5th International Conference on Agro-Geoinformatics, Agro-Geoinformatics 2016, Tianjin, China, 18–20 July 2016.
15. Ticehurst, C.; Guerschman, J.; Chen, Y. The Strengths and Limitations in Using the Daily MODIS Open Water Likelihood Algorithm for Identifying Flood Events. *Remote Sens.* **2014**, *6*, 11791–11809. [CrossRef]
16. Yang, X.; Qin, Q.; Grussenmeyer, P.; Koehl, M. Urban surface water body detection with suppressed built-up noise based on water indices from Sentinel-2 MSI imagery. *Remote Sen. Environ.* **2018**, *219*, 259–270. [CrossRef]
17. Notti, D.; Giordan, D.; Caló, F.; Pepe, A.; Zucca, F.; Galve, J. Potential and Limitations of Open Satellite Data for Flood Mapping. *Remote Sens.* **2018**, *10*, 1673. [CrossRef]
18. Maglione, P. Very High Resolution Optical Satellites: An Overview of the Most Commonly used. *Am. J. Appl. Sci.* **2016**, *13*, 91–99. [CrossRef]
19. Yamazaki, D.; Trigg, M.A.; Ikeshima, D. Development of a global ~90m water body map using multi-temporal Landsat images. *Remote Sens. Environ.* **2015**, *171*, 337–351. [CrossRef]
20. Mueller, N.; Lewis, A.; Roberts, D.; Ring, S.; Melrose, R.; Sixsmith, J.; Lymburner, L.; McIntyre, A.; Tan, P.; Curnow, S.; et al. Water observations from space: Mapping surface water from 25 years of Landsat imagery across Australia. *Remote Sens. Environ.* **2016**, *174*, 341–352. [CrossRef]
21. Tulbure, M.G.; Broich, M. Spatiotemporal dynamic of surface water bodies using Landsat time-series data from 1999 to 2011. *ISPRS J. Photogramm. Remote Sens.* **2013**, *79*, 44–52. [CrossRef]
22. Thy, P.T.M.; Raghavan, V.; Pawar, N.J. Urban expansion of Can Tho City, Vietnam: A study based on multi-temporal satellite images. *Geoinformatics* **2010**, *21*, 13.
23. Mountrakis, G.; Im, J.; Ogole, C. Support vector machines in remote sensing: A review. *ISPRS J. Photogramm. Remote Sens.* **2011**, *66*, 247–259. [CrossRef]
24. Nandi, I.; Srivastava, P.K.; Shah, K. Floodplain Mapping through Support Vector Machine and Optical/Infrared Images from Landsat 8 OLI/TIRS Sensors: Case Study from Varanasi. *Water Resour. Manag.* **2017**, *31*, 1157–1171. [CrossRef]
25. Roli, F.; Fumera, G. Support Vector Machines for Remote-Sensing Image Classification. In Proceedings of the Europto Remote Sensing, Barcelona, Spain, 25–27 September 2001; pp. 160–166.

26. Hoang, L.P.; Biesbroek, R.; Tri, V.P.D.; Kummu, M.; van Vliet, M.T.H.; Leemans, R.; Kabat, P.; Ludwig, F. Managing flood risks in the Mekong Delta: How to address emerging challenges under climate change and socioeconomic developments. *Ambio* **2018**, *47*, 635–649. [CrossRef] [PubMed]
27. Balica, S.; Dinh, Q.; Popescu, I.; Vo, T.Q.; Pham, D.Q. Flood impact in the Mekong Delta, Vietnam. *J. Maps* **2013**, *10*, 257–268. [CrossRef]
28. Ninh, N.H.; Trung, V.K.; Niem, N.X. Flooding in Mekong River Delta, Vietnam. *Hum. Dev. Rep.* **2007**, *2008*, 23.
29. OCHA-ROAP; UNEP/GRID-Europe. Flooding Risk in Asia-Pacific. Available online: https://www.preventionweb.net/files/23470_ocharoapfloodsv6110501.pdf (accessed on 8 July 2019).
30. Neumann, L.; Nguyen, M.; Moglia, M.; Cook, S.; Lipkin, F. *Urban Water Systems in Can Tho, Vietnam: Understanding the Current Context for Climate Change Adaption Climate Adaptation through Sustainable Urban Development*; CSIRO Land & Water: Highett, Australia, 2011.
31. Moglia, M.; Neumann, L.E.; Alexander, K.S.; Nguyen, M.N.; Sharma, A.K.; Cook, S.; Trung, N.H.; Tuan, D.D.A. Application of the Water Needs Index: Can Tho City, Mekong Delta, Vietnam. *J. Hydrol.* **2012**, *468–469*, 203–212. [CrossRef]
32. Central Population and Housing Census Steering Committee VietNam. *Selected Key Indicators The Vietnam Population and Housing Census, 00:00 Hours on 1st April 2019*; Central Population and Housing Census Steering Committee VietNam: Ha Noi, VietNam, 2019.
33. Office of the People's Committee of Can Tho City. Can Tho Portal. Available online: https://www.cantho.gov.vn/wps/portal/home/en/Tourism/!ut/p/z1/04_Sj9CPykssy0xPLMnMz0vMAfIjo8ziLQy8XQ29TIx8DCwcDQwCLcLcPD0c3Y0MXIz0C7IdFQGsFvtY/ (accessed on 24 April 2020).
34. Dang, T.D.; Cochrane, T.A.; Arias, M.E.; Van, P.D.T.; de Vries, T.T. Hydrological alterations from water infrastructure development in the Mekong floodplains. *Hydrol. Process.* **2016**, *30*, 3824–3838. [CrossRef]
35. Ling, F.H.; Tamura, M.; Yasuhara, K.; Ajima, K.; Trinh, C.V. Reducing flood risks in rural households: Survey of perception and adaptation in the Mekong delta. *Clim. Chang.* **2015**, *132*, 209–222. [CrossRef]
36. Triet, N.V.K.; Dung, N.V.; Fujii, H.; Kummu, M.; Merz, B.; Apel, H. Has dyke development in the Vietnamese Mekong Delta shifted flood hazard downstream? *Hydrol. Earth Syst. Sci.* **2017**, *21*, 3991–4010. [CrossRef]
37. Hecht, J.S.; Lacombe, G.; Arias, M.E.; Dang, T.D.; Piman, T. Hydropower dams of the Mekong River basin: A review of their hydrological impacts. *J. Hydrol.* **2019**, *568*, 285–300. [CrossRef]
38. Huong, H.T.L.; Pathirana, A. Urbanization and climate change impacts on future urban flooding in Can Tho city, Vietnam. *Hydrol. Earth Syst. Sci.* **2013**, *17*, 379–394. [CrossRef]
39. Danh, V. *Household Economic Losses of Urban Flooding: Case of Can Tho City, Vietnam*; Southeast Asia Review of Economics and Business: London, UK, 2014; p. 40.
40. Zhang, F.; Zhu, X.; Liu, D. Blending MODIS and Landsat images for urban flood mapping. *Int. J. Remote. Sens.* **2014**, *35*, 3237–3253. [CrossRef]
41. Qi, S.; Brown, D.G.; Tian, Q.; Jiang, L.; Zhao, T.; Bergen, K.M. Inundation Extent and Flood Frequency Mapping Using LANDSAT Imagery and Digital Elevation Models. *GIScience Remote Sens.* **2013**, *46*, 101–127. [CrossRef]
42. Nguyen, U.N.T.; Pham, L.T.H.; Dang, T.D. An automatic water detection approach using Landsat 8 OLI and Google Earth Engine cloud computing to map lakes and reservoirs in New Zealand. *Environ. Monit. Assess.* **2019**, *191*, 235. [CrossRef]
43. Kumar, L.; Mutanga, O. Google Earth Engine Applications Since Inception: Usage, Trends, and Potential. *Remote Sens.* **2018**, *10*, 1509. [CrossRef]
44. Gorelick, N.; Hancher, M.; Dixon, M.; Ilyushchenko, S.; Thau, D.; Moore, R. Google Earth Engine: Planetary-scale geospatial analysis for everyone. *Remote Sens. Environ.* **2017**, *202*, 18–27. [CrossRef]
45. Phongsapan, K.; Chishtie, F.; Poortinga, A.; Bhandari, B.; Meechaiya, C.; Kunlamai, T.; Aung, K.S.; Saah, D.; Anderson, E.; Markert, K.; et al. Operational Flood Risk Index Mapping for Disaster Risk Reduction Using Earth Observations and Cloud Computing Technologies: A Case Study on Myanmar. *Front. Environ. Sci.* **2019**, *7*. [CrossRef]
46. Uddin; Matin; Meyer. Operational Flood Mapping Using Multi-Temporal Sentinel-1 SAR Images: A Case Study from Bangladesh. *Remote Sens.* **2019**, *11*, 1581. [CrossRef]

47. Sidhu, N.; Pebesma, E.; Câmara, G. Using Google Earth Engine to detect land cover change: Singapore as a use case. *Eur. J. Remote. Sens.* **2018**, *51*, 486–500. [CrossRef]
48. Celik, N. Change Detection of Urban Areas in Ankara through Google Earth Engine. In Proceedings of the 41st International Conference on Telecommunications and Signal Processing (TSP), Athens, Greece, 4–6 July 2018; pp. 1–5.
49. Nguyen, N.; Nguyen, M.; Trung, N. Application of GIS and Remote Sensing for assessing changes of built-up areas in Can Tho City from 1988 to 2018. In Proceedings of the ESRI International Conference 2019, Ho Chi Minh City, Vietnam, 15 November 2019.
50. Goldblatt, R.; You, W.; Hanson, G.; Khandelwal, A. Detecting the Boundaries of Urban Areas in India: A Dataset for Pixel-Based Image Classification in Google Earth Engine. *Remote Sens.* **2016**, *8*, 634. [CrossRef]
51. Gao, B.-C. NDWI—A normalized difference water index for remote sensing of vegetation liquid water from space. *Remote Sens. Environ.* **1996**, *58*, 257–266. [CrossRef]
52. McFeeters, S.K. The use of the Normalized Difference Water Index (NDWI) in the delineation of open water features. *Int. J. Remote. Sens.* **1996**, *17*, 1425–1432. [CrossRef]
53. Acharya, T.D.; Subedi, A.; Lee, D.H. Evaluation of Water Indices for Surface Water Extraction in a Landsat 8 Scene of Nepal. *Sensors* **2018**, *18*. [CrossRef] [PubMed]
54. Ahamed, A.; Bolten, J.D. A MODIS-based automated flood monitoring system for southeast asia. *Int. J. Appl. Earth Obs. Geoinf.* **2017**, *61*, 104–117. [CrossRef]
55. Claverie, M.; Ju, J.; Masek, J.G.; Dungan, J.L.; Vermote, E.F.; Roger, J.-C.; Skakun, S.V.; Justice, C. The Harmonized Landsat and Sentinel-2 surface reflectance data set. *Remote Sens. Environ.* **2018**, *219*, 145–161. [CrossRef]
56. Hossain, M.S.; Bujang, J.S.; Zakaria, M.H.; Hashim, M. Assessment of Landsat 7 Scan Line Corrector-off data gap-filling methods for seagrass distribution mapping. *Int. J. Remote. Sens.* **2015**, *36*, 1188–1215. [CrossRef]
57. Chen, F.; Zhao, X.; Ye, H. Making Use of the Landsat 7 SLC-off ETM+ Image Through Different Recovering Approaches. *Data Acquis. Appl.* **2012**. [CrossRef]
58. Dhanotia, R.; Singh, S. A Survey of Image Classification Techniques for Flood Monitoring System. In Proceedings of the International Conference on Emerging Trends in Computer and Image Processing (ICETCIP'2014), Pattaya, Thailand, 15–16 December 2014.
59. Suykens, J.A.K.; Vandewalle, J. Least Squares Support Vector Machine Classifiers. *Neural Process. Lett.* **1999**, *9*, 293–300. [CrossRef]
60. Ireland, G.; Volpi, M.; Petropoulos, G. Examining the Capability of Supervised Machine Learning Classifiers in Extracting Flooded Areas from Landsat TM Imagery: A Case Study from a Mediterranean Flood. *Remote Sens.* **2015**, *7*, 3372–3399. [CrossRef]
61. Syifa, M.; Park, S.J.; Achmad, A.R.; Lee, C.-W.; Eom, J. Flood Mapping Using Remote Sensing Imagery and Artificial Intelligence Techniques: A Case Study in Brumadinho, Brazil. *J. Coast. Res.* **2019**, *90*, 197–204. [CrossRef]
62. Gizaw, M.S.; Gan, T.Y. Regional Flood Frequency Analysis using Support Vector Regression under historical and future climate. *J. Hydrol.* **2016**, *538*, 387–398. [CrossRef]
63. Chen, S.-T.; Yu, P.-S. Real-time probabilistic forecasting of flood stages. *J. Hydrol.* **2007**, *340*, 63–77. [CrossRef]
64. Mosavi, A.; Ozturk, P.; Chau, K.-w. Flood Prediction Using Machine Learning Models: Literature Review. *Water* **2018**, *10*, 1536. [CrossRef]
65. Granata, F.; Gargano, R.; Marinis, G.d. Support Vector Regression for Rainfall-Runoff Modeling in Urban Drainage: A Comparison with the EPA's Storm Water Management Model. *Water* **2016**, *8*, 69.
66. Davison, A.C.; Hinkley, D.V. *Bootstrap Methods and their Application*; Cambridge University Press: Cambridge, UK, 1997. [CrossRef]
67. Meyer, D.; Dimitriadou, E.; Hornik, K.; Weingessel, A.; Leisch, F. *e1071: Misc Functions of the Department of Statistics, Probability Theory Group (Formerly: E1071)*; TU Wien: Vienna, Austria, 2019.
68. R Core Team. *R: A Language and Environment for Statistical Computing*; R Foundation for Statistical Computing: Vienna, Austria, 2019.
69. Boot: Bootstrap R (S-Plus) Functions. R package version 1.3-25. Available online: https://cran.r-project.org/web/packages/boot/boot.pdf (accessed on 28 May 2020).
70. Coltin, B.; McMichael, S.; Smith, T.; Fong, T. Automatic boosted flood mapping from satellite data. *Int. J. Remote. Sens.* **2016**, *37*, 993–1015. [CrossRef]

71. Apel, H.; Merz, B.; Thieken, A.H. Quantification of uncertainties in flood risk assessments. *Int. Int. J. River Basin Manag.* **2008**, *6*, 149–162. [CrossRef]
72. Thy, P.T.M.; Raghavan, V. Monitoring the effect of land cover change on urban inundation by remote sensing and GIS technique in Can Tho City, Vietnam. In Proceedings of the 34th Asian Conference on Remote Sensing (ACRS 2013), Bali, Indonesia, 20–24 October 2013; Asian Association of Remote Sensing: Tokyo, Japan, 2013; pp. 1806–1815.
73. Apel, H.; Trepat, O.M.; Hung, N.N.; Chinh, D.T.; Merz, B.; Dung, N.V. Combined fluvial and pluvial urban flood hazard analysis: Method development and application to Can Tho City, Mekong Delta, Vietnam. *Nat. Hazards Earth Syst. Sci. Discuss.* **2015**, *3*, 4967–5013. [CrossRef]
74. Dang, T.D.; Cochrane, T.A.; Arias, M.E.; Tri, V.P.D. Future hydrological alterations in the Mekong Delta under the impact of water resources development, land subsidence and sea level rise. *J. Hydrol. Reg. Stud.* **2018**, *15*, 119–133. [CrossRef]
75. Eilander, D.; Trambauer, P.; Wagemaker, J.; van Loenen, A. Harvesting Social Media for Generation of Near Real-time Flood Maps. *Procedia Eng.* **2016**, *154*, 176–183. [CrossRef]
76. See, L. A Review of Citizen Science and Crowdsourcing in Applications of Pluvial Flooding. *Frontiers in Earth Science* **2019**, *7*. [CrossRef]
77. McCrory, G.; Veeckman, C. *D1.1 FloodCitiSense Conceptual and Methodological Framework*; Vrije Universiteit Brussel: Brussels, Belgium, 2017. [CrossRef]
78. Hultquist, C.; Cervone, G. *Integration of Crowdsourced Images, USGS Networks, Remote Sensing, and a Model to Assess Flood Depth during Hurricane Florence*; MDPI AG: Basel, Switzerland, 2020; Volume 12, p. 834.
79. Tehrany, M.S.; Pradhan, B.; Jebur, M.N. Spatial prediction of flood susceptible areas using rule based decision tree (DT) and a novel ensemble bivariate and multivariate statistical models in GIS. *J. Hydrol.* **2013**, *504*, 69–79. [CrossRef]
80. Liew, S.C.; Gupta, A.; Chia, A.S.; Ang, W.C. The flood of 2011 in the lower Chao Phraya valley, Thailand: Study of a long-duration flood through satellite images. *Geomorphology* **2016**, *262*, 112–122. [CrossRef]

*water*

MDPI

*Article*

# An Integrated Approach for Assessing Flood Risk in Historic City Centres

**Tiago M. Ferreira [1,*] and Pedro P. Santos [2]**

1 ISISE, Institute of Science and Innovation for Bio-Sustainability (IB-S), Department of Civil Engineering, University of Minho, 4800-058 Guimarães, Portugal

2 Centre for Geographical Studies, Institute of Geography and Spatial Planning, University of Lisbon, 1600-276 Lisbon, Portugal; pmpsantos@campus.ul.pt

* Correspondence: tmferreira@civil.uminho.pt

Received: 9 May 2020; Accepted: 7 June 2020; Published: 9 June 2020

**Abstract:** Historic city centres near watercourses are a specific type of urban area that are particularly vulnerable to flooding. In this study, we present a new methodology of flood risk assessment that crosses hazard and physical vulnerability information. We have selected the Historic City Centre of Guimarães (Portugal), a UNESCO Heritage Site, for developing and testing the defined methodology. The flood hazard scenario was obtained through the hydrologic–hydraulic modelling of peak flows with a 100-year return period, which provided flood extent, depths, and velocities. A decomposition of the momentum equation, using depth and velocity, allowed reaching a final hazard score. Flood vulnerability was assessed through combining an exposure component and a sensitivity component, from field-collected data regarding wall orientation, heritage status, age, number of storeys, condition, and material of buildings. By combining the results of the hazard and vulnerability modules in a risk-matrix, three qualitative levels of flood risk were defined. The individual and crossed analysis of results proved to be complementary. On one hand, it allows the identification of the more relevant risk factors—from the hazard or vulnerability modules. On the other hand, the risk-matrix identified other buildings with a high risk that otherwise would remain unnoticed to risk managers.

**Keywords:** urban flood; flood risk assessment; risk management; Historic City Centre of Guimarães

## 1. Introduction

The dizzying rate of urbanisation over the last few decades has led to an exponential increase in the magnitude of losses caused by natural and technological hazards worldwide. According to the risk-related literature, these losses can be understood as the consequence of a certain level of risk associated with a specific community or society over some specified time period [1]. In a broader sense, disaster risk is a compound concept determined by the combination of the hazard, exposure, and intrinsic vulnerability. Moreover, because vulnerability and hazard may change over time, disaster risk is highly dynamic [2,3]. For this reason, it is fundamental to have ways of monitoring this risk and, when necessary, supporting the definition and implementation of risk mitigation actions.

As a result of their heritage value and high physical vulnerability, historical centres are particularly critical areas and, therefore, deserve particular attention. According to the literature [4,5], in the last few decades, the impact of flooding events in historic centres has indeed increased steadily. As reported by Miranda and Ferreira [6], examples of recent high-impacting flooding events in historical centres include those that occurred in Central Europe, back in 2002 and 2010, in South Asia, in 2007 and 2008, and the New Orleans flood of 2005, caused by Hurricane Katrina. In a recent attempt to study this phenomenon, Marzeion and Levermann [7] investigated the number of cultural heritage sites worldwide that, due to global warming, are at risk of being flooded in the next two millennia.

The results of this study are quite conclusive, pointing out that approximately 6% of current UNESCO sites (about 40 sites) will be flooded, particularly in China and India.

The present article addresses this challenge by discussing the application of an integrated flood risk assessment approach, which combines flood hazard and building vulnerability indicators to identify and classify risk and to narrow intervention priorities. The Historic City Centre of Guimarães (Portugal)—a World Heritage Site inscribed by UNESCO since 2001 due to its extraordinary authenticity and well-preserved condition—is explored here to illustrate the application of the methodology. After modelling the flood hazard using the hydrologic–hydraulic method and evaluating the flood vulnerability of the buildings by resorting to a simplified vulnerability assessment method, we provide a comprehensive analysis of the outputs, both in an individual and integrated manner. Finally, we used a risk-matrix approach to aggregate these hazard and vulnerability outputs and categorise the buildings into three qualitative levels of risk.

More than the results themselves—which are indeed very important for the Municipality of Guimarães, but eventually will be of little relevance to the reader—the interest and the innovation of this paper lies in how this integrated flood risk assessment approach (which results from the original combination of two already existing methods) can be used to individualise and guide intervention decisions. It is this analysis that, we believe, can be of great interest and direct relevance to the reader's own practice.

## 2. The Historic City Centre of Guimarães

Located in northern Portugal, the city of Guimarães is usually referred to as the cradle of the Portuguese nationality [8]. The Historic City Centre of Guimarães is a World Heritage Site, deemed as such in 2001. According to the classification committee, the Historic City Centre of Guimarães is an exceptionally well-preserved and authentic example of the evolution of a Medieval settlement into a modern town, see Figure 1a. The architectural characteristics of the Historic City Centre of Guimarães, marked by a rich diversity of construction typologies associated with different evolution periods of the place, as well as its unity and integration with the landscape setting, compounds outstanding universal values.

(a)

(b)

**Figure 1.** The Historic City Centre of Guimarães: (**a**) View to one of its squares; (**b**) Example of a flood event in the Couros' river basin (source: Guimarães City Council).

According to Miranda and Ferreira [6], in recent decades, the city of Guimarães has been subjected to intense anthropogenic pressure as a result of a steady increase of urban and industrial occupation. This has led to the depreciation of the Couros river basin, which was once a central part of the population's life and a vital element for the development of the leather industry [9]. As a result, the levels of pollution and contamination in the Couros creek, as well as the severity of flood events in the Historic City Centre of Guimarães, have increased substantially, Figure 1b.

The coexistence of the aspects mentioned above makes the assessment of the flood risk in the Historic City Centre of Guimarães a particularly challenging and relevant work, constituting, therefore, the justification for the selection of this case study.

*Past Flood Events in the Historic City Centre of Guimarães*

The Couros river basin (11.2 $km^2$) is part of the Selho river basin (67.7 $km^2$), which is a sub-basin of the Ave river basin (1390 $km^2$) which drains to the Atlantic Ocean. The Couros river (5.6 km length) is thus a small left tributary of Selho river, located entirely in the Municipality of Guimarães.

Water availability soon became a relevant driver of human occupancy along the Couros river valley that crosses the Historic City Centre of Guimarães.

According to the DISASTER database [10], there are 10 flood occurrences in the municipality. This database covers the period from 1865 to 2015 and includes only the most severe cases of flooding (i.e., when at least one casualty, injured, missing, displaced, or evacuated person is registered in the data sources). In the study area, along the Couros river, there are two flood occurrences of the DISASTER type, both occurring during the flood event of 28 February 1978: in the elder residential structure "Lar dos Santos Passos", where 23 persons were evacuated; and a few hundred meters downstream, at the trucking station, where three families were evacuated.

Despite this apparently modest record of severe losses, the dense artificialisation and imperviousness of the basin—particularly in the Historic City Centre—has been causing frequent episodes of flooding with only material consequences and economic losses. The most recent occurred on 20 December 2019, which was associated with the Elsa atmospheric depression.

## 3. Materials and Methods

To better explain the methodological framework adopted in this work, this section presents and discusses the fundamentals of the applied flood risk assessment method. As schematically illustrated in Figure 2, the approach is composed of two main modules, the hazard and the vulnerability module. The bases of these two modules are detailed in Sections 3.1 and 3.2. Finally, Section 3.4 introduces the study area and the approach adopted in this research to collect, manage, and explore data.

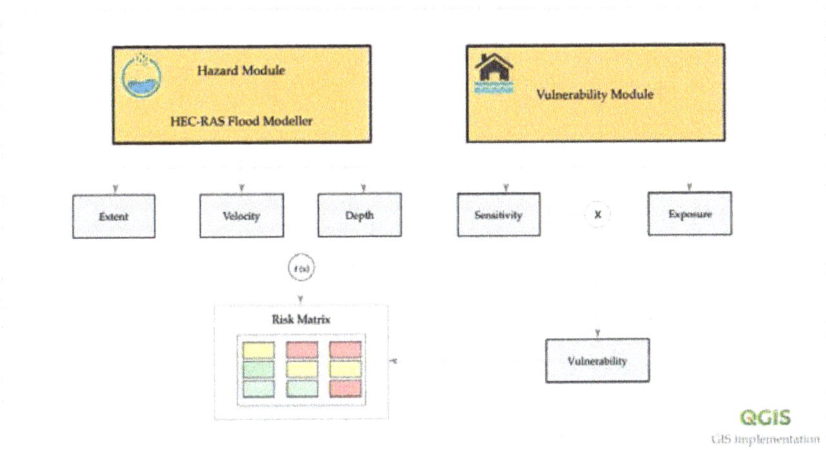


**Figure 2.** The conceptual framework of the integrated flood risk assessment approach.

### 3.1. Hazard Module

The flood hazard was assessed using the hydrologic–hydraulic method. The assessment process involved the acquisition and preparation of geometric data, the estimation of the peak flow, the hydraulic modelling, and the GIS post-processing and mapping.

In the first step, we obtained the geometric data, representing the morphological features of the floodplain from a base map at a scale of 1:1000, contour lines 1 m equidistant, and a dense cover of mass points. Such data allowed us to create a detailed digital elevation model (DEM) with 1 m cell size, covering the entire valley of the Couros river, oriented E-W and crossing the Historic City Centre of Guimarães, illustrating how significantly modified by human intervention the natural morphology of the study area has been. The DEM features a topographic obstruction crossing the floodplain– and the Couros river—with an orientation N-S. At this section, flow occurs through a hydraulic passage of 30 m length, with a rectangular shape of 4.0 m × 3.5 m, which was geometric and hydraulically modelled as a bridge. Roughness was represented through the Manning's $n$ value: 0.025 in the channel and 0.05 in overbank areas.

Flow data were estimated for the 100-year flood, based on the results obtained from the hydrologic study of Ramísio, Duarte, and Vieira [11,12], in which peak flows were obtained using empirical, kinematic, and statistical methods. The adopted 100-year peak flow results from a simple average of the estimates presented in the cited study, regarding four methods. Three of the methods are kinematic: the Giandotti, the rational method (using Kirpich concentration-time), and the rational method (using the Chow concentration-time). As for the rational method, we used the rainfall Intensity-Duration-Frequency (IDF) curve parameters from a rain gauge station located 45 km ENE of the study area, with a time series of 25 years. Although the gauge is located 45 km from the study area, it represents the same climatological and pluviogenic context found at the Couros river basin. Although the rain gauge is located at an elevation of 95 m, its 10 km radius has a mean elevation of 331.4 m, which is circa 70 m above the mean basin's elevation of 258.8 m. Considering the mountainous context where both areas are located, such a difference can be considered as not relevant. Of the several rainfall durations for which those parameters are valid, we have adopted the interval 30 min to 6 h, which frames the concentration-time found in the upstream sub-basin that drains to the modelled reach—30.8 min and 71.7 min according to the Kirpich and Chow formulas, respectively. The rainfall duration is assumed to equal the concentration-time obtained from the simple average of the applied methods. The runoff coefficient C was estimated from the slope angle, land use data, and the degree of imperviousness [11]: 0.56 in the upstream basin and 0.73 in the basin crossed by the modelled stream of the Couros river.

The fourth method is a statistical one, designated as the Loureiro's formula. This method uses the area of the basin and regional parameters, valid for the Portuguese context only, calculated from peak flow data fitted to the Gumbel distribution. The average of the four estimations was used from the Ramísio, Duarte, and Vieira [12] study in two sections: one, defined at the upstream inlet (35.95 $m^3/s$, corresponding to a contributing basin of 3.75 $km^2$); the other, estimated at the downstream outlet of the modelled reach (54.15 $m^3/s$), the difference (18.2 $m^3/s$) of which we have distributed along the 1786 m length of the modelled reach, proportionally to the drainage area of 11 pour points, Table 1.

**Table 1.** Estimated flow data for use in the hydraulic model of the Couros river reach, in the Historic City Centre of Guimarães.

| Pour Point | Upstream Inlet | A | B | C | D | E | F | G | H | I | J | K |
|---|---|---|---|---|---|---|---|---|---|---|---|---|
| Reach length (m) | 0.00 | 33.0 | 51.6 | 158.9 | 260.0 | 651.2 | 782.7 | 1013.7 | 1146.4 | 1252.6 | 1344.2 | 1495.1 |
| Affluent Qp ($m^3/s$) | 35.95 | 0.45 | 1.53 | 0.60 | 2.00 | 1.16 | 0.32 | 1.97 | 7.55 | 0.38 | 0.97 | 1.27 |
| Sum Qp ($m^3/s$) | 35.95 | 36.40 | 37.93 | 38.53 | 40.53 | 41.69 | 42.01 | 43.98 | 51.53 | 51.91 | 52.88 | 54.15 |

By adopting this approach, we have intended to consider the effect of the complex sewer system that drains the small urban basins directly to the Couros river along the modelled reach.

Pre-processing of geometric data was performed at the HEC-RAS 5.0.7 environment [13], using the RAS Mapper tool. The modelled reach goes further downstream from the area to which the vulnerability of buildings was available. This is explained by the existence of the mentioned obstruction near the study area, leading us to model a flood pathway long enough to allow a proper elongation of floodwaters downstream of the obstruction. A normal depth slope of 0.015 m/m was used to define the downstream reach boundary conditions. A steady flow water surface computation was initiated at the upstream boundary using the 35.95 $m^3/s$ peak flow, to which the mentioned 11 flow change locations were sequentially added. In the modelling plan, we selected a mixed flow regime and requested the optional floodplain mapping.

Finally, we sequentially exported the HEC-RAS modelling results to SDF, XML, and GIS formats. Velocity and depth data were exported as GRID files (ESRI raster format) and the flood boundary as shapefile (vector polygon), both for the 100-year flood.

### 3.2. Vulnerability Module

Flood vulnerability is assessed through the application of the simplified flood vulnerability assessment methodology, which was initially proposed by Miranda and Ferreira [6]. This methodology consists of the evaluation of two vulnerability components, an exposure and a sensitivity component (Figure 2), which, through an index, quantify the vulnerability of the building to flood inundation.

As presented in Figure 3, the Exposure Component is composed of one single parameter (Wall Orientation), which evaluates the influence of the orientation of the main façade wall of the building to the water flow. This parameter intends to bring together the multiple aspects behind this phenomenon: the location of the building, the orientation of its main façade wall, and the existence of openings, knowing that buildings located in low-lying areas are theoretically more susceptible to inundation due to runoff. Regarding the Sensitivity Component, this focusses on the physical characteristics of the building by evaluating its Material characteristics, Condition (or conservation state), Number of Storeys, Age, and Heritage Status. It is worth noting that these indicators were defined from a comprehensive review of analogous indicators proposed for assessing similar building typologies and structural characteristics under similar conditions [6].

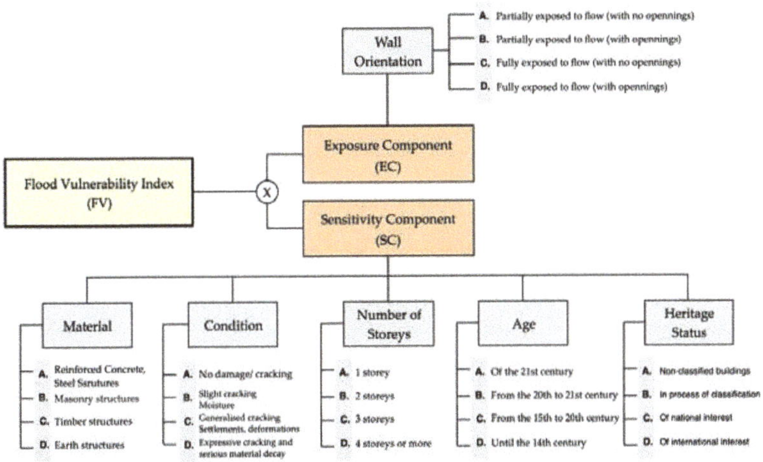


**Figure 3.** The schematisation of the vulnerability module.

We obtained the values of the exposure (EC) and the sensitivity (SC) components, which was done based on the vulnerability classes identified in Figure 3 (where A = 10, B = 40, C = 70, and D = 100), and individual flood vulnerability (FV) can be obtained using Equation (1).

$$FV = EC \times SC. \tag{1}$$

For simplicity's sake, the values of these three indices (EC, SC, and FV)—which range, respectively, from 10 to 100, 50 to 500, and 500 to 50,000—are normalised to fall within the range of 0 to 100; the lower the value of the index, the lower the level of exposure, sensitivity, or vulnerability of the building. According to Miranda and Ferreira [6], despite the fact that this approach was primarily intended for assessing the flood vulnerability of single buildings, its simple formulation makes it particularly suitable to evaluate large urban areas, such as historic city centres. It is precisely in this context that we apply this methodology herein.

### 3.3. Risk Matrix

Finally, flood risk is computed from the combination of the hazard and vulnerability results obtained by using the above-presented approaches. This is done through a vulnerability–hazard matrix, which relates each building's vulnerability with the level of hazard to which it is exposed (see Table 2).

**Table 2.** Flood risk matrix.

| **Flood Risk** | | **Hazard** | | |
|---|---|---|---|---|
| | | Low | Moderate | High |
| **Vulnerability** | High | Middle Risk | High Risk | High Risk |
| | Moderate | Low Risk | Middle Risk | High Risk |
| | Low | Low Risk | Low Risk | Middle Risk |

Regarding the vulnerability level, this is measured directly by the flood vulnerability index. For such, 20 and 40 were conservatively defined here as plausible threshold values for "moderate" and "high" vulnerability, respectively. Although this criterion may be debatable, it is important to note that the value of 40 (the boundary value between "moderate" and "high") is frequently used in index-based vulnerability assessment approaches as a threshold for high vulnerability, see, for example, [14]. Concerning the level of hazard, this is obtained by combining flood velocity ($v$) and water depth ($y$) results according to the criterion given in Equation (2):

$$\begin{cases} v < 2\,m/s \vee y \times v < 3\,m^2/s \triangleq \text{Low Hazard} \\ v > 2\,m/s \bigwedge 3\,m^2/s < y \times v < 7\,m^2/s \triangleq \text{Moderate Hazard} \\ v > 2\,m/s \bigwedge y \times v > 7\,m^2/s \triangleq \text{High Hazard.} \end{cases} \tag{2}$$

This criterion was proposed originally by Clausen [15] based on empirical data. According to Kelan and Spence [16], a physical meaning for this condition is related to momentum $=$ mass $\times v =$ $p_w \times$ volume $\times v = p_w \times$ horizontal flood area $\times y \times v$. If $p_w$ is constant in this hypothesis, the horizontal flood area can be considered constant, thus leaving $y \times v$ as the variable.

### 3.4. Study Area, Building Assessment, and Implementation of a Geographic Information System (GIS) Tool

The study area includes nine blocks with 116 buildings in total. Each one of these buildings was comprehensively assessed onsite to collect the data required for applying the flood vulnerability assessment approach, by using a detailed checklist. After being digitalised and systematised, the data that were contained in the checklists were manually inputted into a spreadsheet database to create a digital record and to automate some later steps of the work.

Once all the indices and flood indicators were computed, the results were plotted and analysed spatially using the open-source Geographic Information System software QGIS [17]. Geo-referenced graphical data (i.e., vectorised information and orthophoto maps) and specific information related to the hazard and the characteristics of the buildings were combined within the software to obtain first- and second-order outputs. In this case, each polygon (corresponding to a building) is associated with several features and attributes, allowing for their visualisation, selection, and search.

Because this GIS tool can efficiently combine highly-relevant hazard, vulnerability, and risk outputs in a very flexible and dynamic environment (easily updatable or modified at any time), it is undoubtedly a significant asset for risk management purposes, allowing the local authorities to define more consequent risk mitigation strategies.

## 4. Results and Discussion

Before getting into the integrated risk assessment outputs in Section 4.3, it is worth exploring the individual hazard and vulnerability results (in Sections 4.1 and 4.2, respectively), which we define as first-order results.

### *4.1. Hazard Modelling*

The flood modelling process described in Section 3.1 allowed us to obtain a broad set of primary hazard indicators, which, alone, give us good insight into the potential magnitude of a flood event in the study area. As already noticed in Section 3, we focused on three primary hazard indicators: flood extent, velocity, and water depth.

As for the flood extent, we found that for the adopted 100-year peak flow scenario, the study area was significantly affected by the flood. As illustrated in Figure 4, 41 out of the 116 buildings considered in this analysis are potentially affected by the flood.

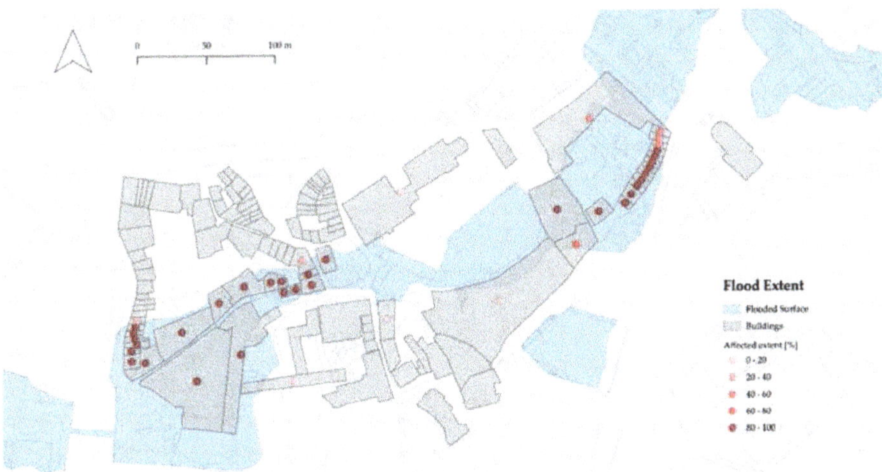


**Figure 4.** Flood inundation map for the adopted 100-year peak flow scenario.

Of these 41 buildings, 23 are affected to the fullest extent, whereas the other 18 are partially flooded, see Table 3. In absolute numbers, about 11,000 m$^2$ of a total of about 32,000 m$^2$ of built-up area are affected, which corresponds to about 34%.

**Table 3.** Number of affected building distributed by ranges of the affected extent.

| Flood Extent | Range of Values | | | | | |
|---|---|---|---|---|---|---|
| | >0–20 | 20–40 | 40–60 | 60–80 | 80–<100 | Fully affected |
| **Affected Buildings** | 4 (9.76%) | 2 (4.88%) | 4 (9.76%) | 1 (2.44%) | 7 (17.07%) | 23 (56.10%) |

Concerning the flood velocity, presented in Figure 5, it ranges between 0.01 m/s and 5.72 m/s. The average velocity value at the surface of the 41 buildings affected by the flood is about 2.15 m/s—with a standard deviation value (STD) of 1.68—being that 22 of these 41 building are exposed to surface velocities higher than 2 m/s. As can be observed in Figure 5, there are two blocks that are particularly affected: one located in the northeast zone (15 buildings) and the other roughly at the central zone of the study area (7 buildings).

**Figure 5.** Flood velocities resulted from the adopted 100-year peak flow scenario.

As for the water depth, from our hazard analysis, it was found that for the considered 100-year peak flow scenario, the buildings will expectably be exposed to an average depth of about 1.20 m (STD = 0.97). As illustrated in Figure 6, 31 out of the 41 buildings affected—which is slightly more than 75%—present a water height of more than 0.5 m, which is a very significant value. The maximum water depth value was obtained at the triangular shape building located at the southwest zone of the study area (4.58 m), see Figure 6.

Table 4 summarises the above-discussed results, presenting the absolute and relative number of potentially affected buildings for different ranges of flood velocity and water depth.

**Table 4.** Number of affected buildings distributed by ranges of flood velocity and water depth.

| Hazard Indicator | Range of Values | | | | | | | | |
|---|---|---|---|---|---|---|---|---|---|
| | 0–0.5 | 0.5–1.0 | 1.0–1.5 | 1.5–2.0 | 2.0–2.5 | 2.5–3.0 | 3.0–3.5 | 3.5–4.0 | >4.0 |
| Velocity (m/s) | 12 (29.3%) | 3 (7.3%) | 1 (2.4%) | 3 (7.3%) | 5 (12.2%) | 5 (12.2%) | 4 (9.8%) | 3 (7.3%) | 5 (12.2%) |
| Depth (m) | 10 (24.4%) | 15 (36.6%) | 4 (9.8%) | 5 (12.2%) | 2 (4.9%) | 2 (4.9%) | 1 (2.4%) | 1 (2.4%) | 1 (2.4%) |

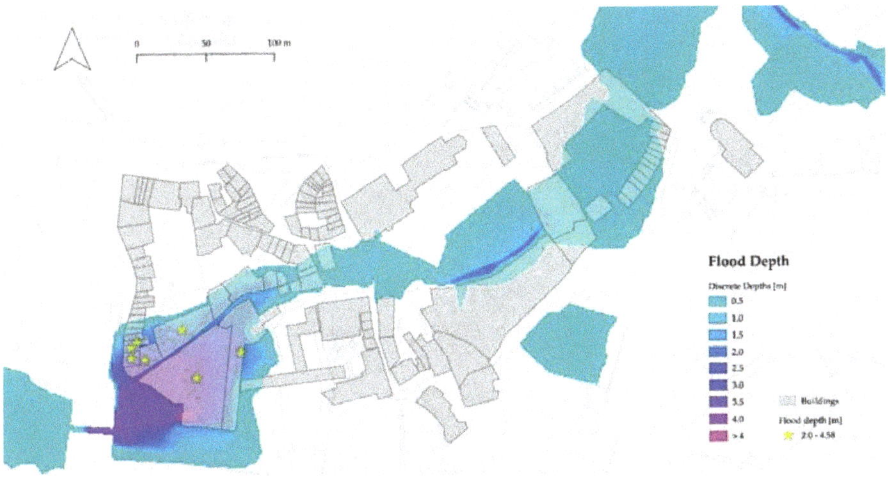


**Figure 6.** Water depth resulted from the adopted 100-year peak flow scenario.

*4.2. Flood Vulnerability*

From the application of the flood vulnerability assessment approach detailed in Section 3.2, we found that 60% of the 116 buildings evaluated present a flood vulnerability value (FV) between 0 and 30. As presented in Figure 7, the remaining 40% present values ranging between 30 and 100. Statistically, the dataset has an average value of 25.70 (STD = 15.95).

Although flood damages cannot be estimated directly from these vulnerability values, the distribution shown in Figure 7 seems to demonstrate that a significant proportion of the buildings assessed are potentially very vulnerable to a flood event. We will provide more insight into this in Section 4.3.

When exploring vulnerability results, it is often relevant to dive into the analysis of the specific parameters of the methodology. Such analysis allows for a better understanding of the vulnerability sources, which is a fundamental prerequisite to defining more consequent and effective risk mitigation measures. Figure 6 presents a set of six maps associated with the spatial distribution of the parameters that compose the flood vulnerability index.

A comprehensive discussion of the maps gathered in Figure 7 would be of little interest. However, we find it relevant to highlight some main results.

Firstly, the general conservation state of the buildings: As it is apparent in Figure 7c, in general, the buildings are either in good condition (Vulnerability Class A, 34%) or have minor conservation issues (Class B, 46%). Most of these minor conservation issues are related to small cracks or coating decay. Further, 20% of the buildings are in poor condition, presenting a significant cracking and moisture phenomena.

Secondly, the distribution of the number of storeys, in Figure 7d: This aspect is especially relevant since, according to several authors [18,19], the number of storeys has a direct influence on the vulnerability of the building to flooding. We will go into further detail regarding this when discussing the second-order risk results. For now, let us emphasise that 113 out of the 116 buildings evaluated have between 1 and 3 floors, distributed as follows: 38% are single-storey buildings, 33% have two storeys, and 27% have three.

Finally, the Heritage Status, in Figure 7f: This is a relevant aspect in the sense that buildings with heritage value deserve particular attention when it comes to risk assessment. Thus, regarding this aspect, 59% of the buildings assessed are ordinary buildings (i.e., non-classified). Then, 38% are currently in the process of classification and only 3% correspond to classified buildings.

**Figure 7.** Mapping of the spatial distribution of the vulnerability classes: (**a**) Wall Orientation; (**b**) Building Material; (**c**) Condition (or conservation status); (**d**) Number of Storeys; (**e**) Building Age; (**f**) Heritage Status.

When getting into the discrete exposure, sensitivity, and vulnerability results, it is possible to gain a much better understanding of the overall vulnerability of the buildings. As illustrated in Figure 8a,b, respectively, spatial analysis reveals a scattered distribution of exposure and the sensitivity indicators over the study area. Comparatively, it is also clear that the exposure values are generally higher than sensitivity values. Despite this, and although there is no correlation between these two indicators, it is interesting to see that some of the most exposed buildings are also those that revealed to be more sensitive. This cross analysis is presented in Figure 9a, where we identify the buildings to which the exposure and the sensitivity values are cumulatively higher than 40. For vulnerability reduction purposes, these buildings are the most critical ones.

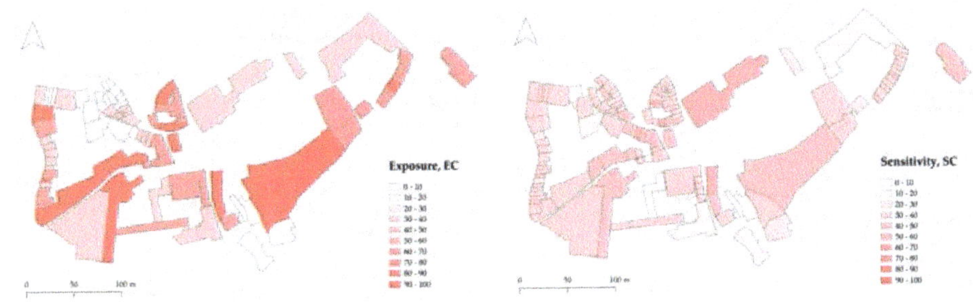


(a) (b)

**Figure 8.** Spatial distribution of the exposure (**a**) and the sensitivity (**b**) indicators.

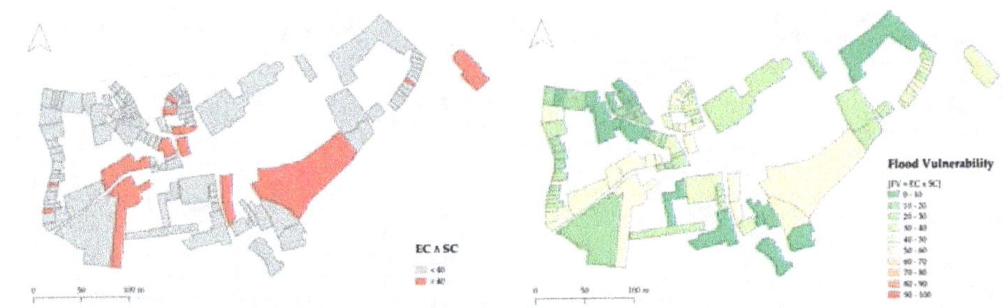


(a) (b)

**Figure 9.** Cross analysis of the exposure and the sensitivity indicators (**a**) and spatial distribution of the vulnerability index results (**b**).

Figure 9b presents the distribution of the vulnerability index results. Although we have already commented on the most significant aspects when addressing the exposure and the sensitivity results, it is worth emphasising two further points. First, the fact that a significant part of the most vulnerable buildings is located in the central part of the study area. As we will have the opportunity to prove afterwards, this aspect may be essential, taking into account that these more vulnerable buildings are located coincidentally within the most hazardous area. Second, the fact that some of these buildings are abandoned. Keeping the social aspects out of the discussion, buildings' abandonment is one of the main factors of rapid degradation and, as a result, increased physical vulnerability. During the discussion of the risk results, we will provide further insight into these critical issues.

### 4.3. *Integrated Risk Assessment*

After having discussed the main outputs resulting from the hazard and the vulnerability analysis, we are in an excellent position to integrate these results in order to obtain more comprehensive flood risk indicators. As detailed in Section 3.3, this integration will ultimately result in a risk matrix that correlates the level of vulnerability with the level of hazard to which each building is exposed. However, before going into such an outcome, we would like to analyse a set of second-order results obtained by crossing some of the above-discussed indicators.

#### 4.3.1. Second-Order Analysis

The first second-order result that we think is of great interest is the joint analysis of the water depth, the exposure component of the vulnerability index, and the conservation state of the buildings. Before going into the result, it is worth justifying the selection of these particular indicators. Water depth is recognised as the most relevant hazard indicator when analysing the impact of flood actions on buildings [15]: depth is usually used to produce vulnerability curves associated with flood events, which is also known as depth-damage curves [20–22]. In fact, although some studies note the importance of flood parameters other than depth [16], those are barely analysed so comprehensively. Equally important is the fact that water depth is directly related to flow velocity and, therefore, with the effects of the hydrodynamic actions. The impact of these hydrodynamic actions on the buildings are, of course, very dependent on their level of exposure—a building of which its façade wall is perpendicular to the direction of the water flow is potentially much more affected than another where its façade is parallel to the direction of the flow. As referred to in Section 3, this is exactly what the sensitivity component of the flood vulnerability index seeks to evaluate and that is why it is considered in this second-order analysis. It is also known that the weaker the state of conservation of the building, the higher the impact of the flood (due to hydrostatic and hydrodynamic actions). This fact rationalises the inclusion of this aspect in the present analysis.

Justifying the three indicators considered herein, Figure 10 presents the map resulting from their integrated analysis. We want to highlight a couple of interesting conclusions from the analysis of this map. The first one is the identification of highly-exposed buildings (i.e., with an EC value higher than 40) that, for the adopted 100-year peak flow scenario, are subject to a water depth over 2 m. Based on this first criterion, it is possible to identify six buildings—about 5% of the building stock within our study area—which are highlighted in red in Figure 10. However, when including the conservation state into the analysis, this number can be further reduced to 5 (about 4% of the building stock), making this result even more informative for decision-making purposes.

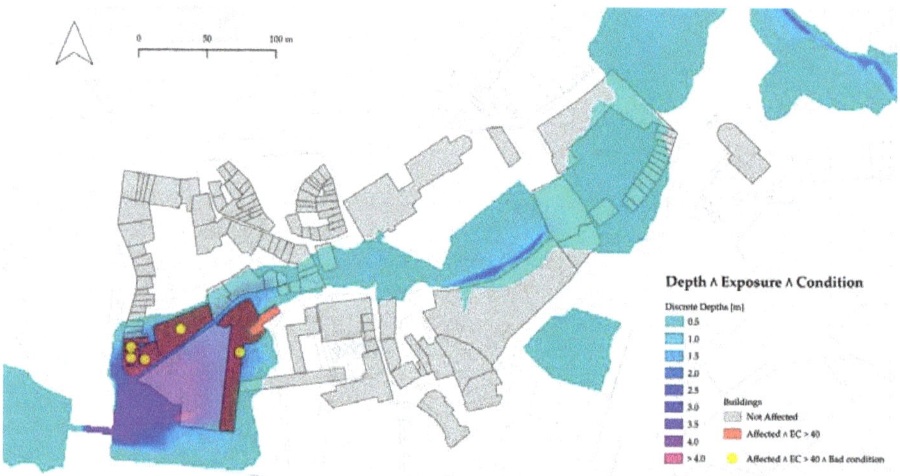


**Figure 10.** Joint analysis of water depth, exposure, and condition indicators.

Another second-order result that we find worthy of particular discussion herein is the joint analysis of the water depth and the sensitivity component of the flood vulnerability index methodology. Since this component assesses the intrinsic characteristics of the buildings that make them sensitive to the impact of flood actions, it is undoubtedly relevant to consider these two indicators together. We would like to note that the conservation sate of the buildings is already part of the sensitivity component, which is why we now choose not to consider this aspect explicitly. Figure 11 presents the maps resulting from this analysis.

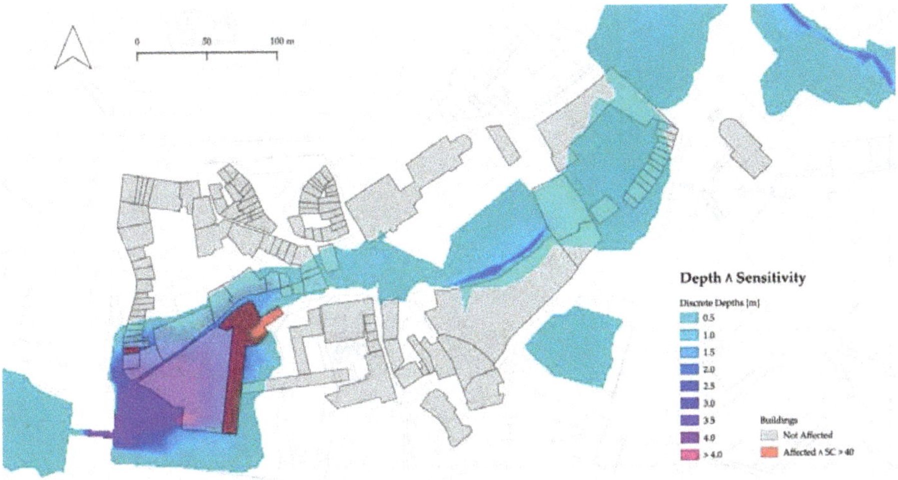


**Figure 11.** Joint analysis of water depth and the sensitivity indicator.

As can be inferred from the analysis of Figure 11, the number of buildings identified from the joint analysis of the water depth and the buildings' sensitivity (for the very same 100-year peak flow scenario) is reduced to 2, which is a little less than 2%. It is also interesting to note that these two buildings are among the five already identified in Figure 10, which, from a decision-making standpoint, pushes them to the top of the intervention priorities.

#### 4.3.2. Matrix-Based Analysis

After the above preliminary second-order analysis—through which we have already gained a deeper understanding about the potential impact of the considered flood scenario in some particular buildings—we got to the point of combining the obtained hazard and vulnerability results into a single flood risk indicator. As a result, the level of hazard and vulnerability associated with each building are related through a risk matrix, wherein the vulnerability is inputted directly using the flood vulnerability index (FV) and the hazard is derived from the velocity and water depth results, using Equation (2).

Thus, before diving deeper into the results obtained from the flood risk matrix, it is useful to map and analyse the spatial distribution of the hazard and vulnerability results. These maps are presented in Figure 12a,b. According to the criterion adopted in this matrix-based analysis, the level of flood hazard throughout the study area is quite homogeneous, see Figure 12a. Further, 22 out of the 116 buildings assessed (about 19%) were identified as having a "Moderate" flood hazard, whereas all the remaining were labelled with "Low" hazard. In terms of spatial distribution, it is possible to identify two blocks that can be particularly affected. Let us notice that although the hazard is being evaluated differently in this section—here, velocity and water depth are combined into a single hazard indicator—this result is in essential agreement with the discussion provided in Section 4.1. As for the spatial distribution of the flood vulnerability results, given in Figure 12b, it is visibly more heterogeneous: 37.9% (44), 41.4% (48), and 20.7% (24) of the buildings were identified as having a "Low", "Moderate", and "High" flood vulnerability, respectively.

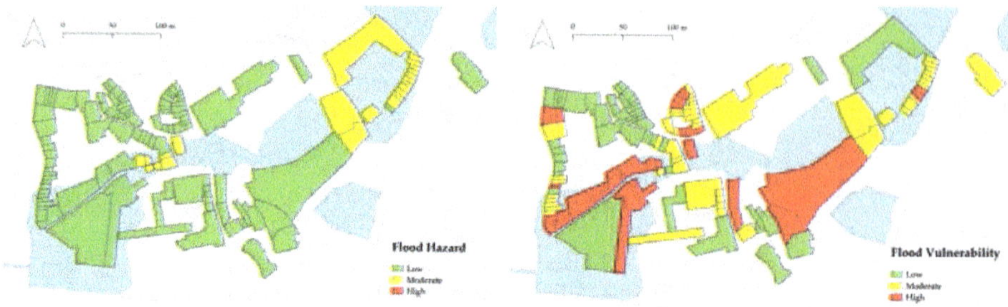


(a) (b)

**Figure 12.** Spatial distribution of the flood hazard (**a**) and flood vulnerability results (**b**).

Contextualising the hazard and the vulnerability results, it remains for us to present and discuss the final flood risk results. They are illustrated in Figure 13 and quantified in Table 3.

**Figure 13.** Spatial distribution of the flood risk results.

As shown in the flood risk matrix provided in Table 5, 66.4% of the buildings (77) were identified with "Low" flood risk, 29.3% (34) with "Middle" risk, and 4.3% (5) with "High" risk for the same 100-year peak flow scenario considered in this work.

**Table 5.** Flood risk matrix.

| **Flood Risk** | | **Hazard** | | |
|---|---|---|---|---|
| | | Low | Moderate | High |
| **Vulnerability** | High | 19 (16.4%)<br>Middle Risk | 5 (4.3%)<br>High Risk | 0<br>High Risk |
| | Moderate | 33 (28.5%)<br>Low Risk | 15 (12.9%)<br>Middle Risk | 0<br>High Risk |
| | Low | 42 (36.2%)<br>Low Risk | 2 (1.7%)<br>Low Risk | 0<br>Middle Risk |

Complementarily to the main conclusions drawn from the second-order analysis discussed in Section 4.3.1, this outcome allows us to recognise some additional buildings that, because the hydrodynamics effects of the flood have been disregarded in that analysis (only water depth was considered explicitly), are not identified there. In fact, none of the building previously highlighted in Figures 10 and 11 are identified in this final analysis as having a "High" flood risk. If in a less thoughtful consideration these two results may seem divergent, in fact, they prove to us the importance of considering different criteria and approaches to assess flood risk in urban areas. Still, in this regard, we would like to stress that these results are obviously conditioned by the criterion used to define the levels of hazard and vulnerability. If, for example, we had used a different criterion from that given by Equation (2)—which, as is always the case, was proposed by Clausen [15] from a set of specific conditions—the results could be very different. This said, the five buildings identified in this analysis, together with those five identified in Section 4.3.1, should be the priority targets of future flood risk mitigation programs.

## 5. Conclusions

This paper presents a new method for assessing the flood risk of the built environment, specifically adapted to areas classified by their heritage value. For the effect, the Historic City Centre of Guimarães (Portugal) was selected as a study area.

We have defined a hazard scenario for the 100-year flood by hydraulic modelling with HEC-RAS, of which its outputs are flood extent, depth, and velocity. Such outputs, derived from a steady flow analysis, must assume the limitation of not providing information that ultimately would allow for a better understanding of the flooding process, namely that from a flood hydrograph (time to peak and duration of inundation), an unsteady analysis would result. The vulnerability has been assessed by considering an exposure component (based on wall orientation) and a sensitivity component (based on heritage status, age, number of storeys, condition, and material of buildings). They define the hazard and vulnerability modules of a GIS tool, which later provided a cross-analysis of information, culminating in a flood risk matrix. A total of 116 buildings were evaluated with the developed methodology.

The first insight into each module's results provided a detailed understanding of flood risk roots or causes, which gives decision-makers and planners information on which risk factors are more relevant in each block or building. This highlighted wall orientation and condition as the most concerning aspects, with exposure being more relevant than sensitivity in explaining physical vulnerability. The second-order analysis evaluated evidence risk contexts that otherwise would go unnoticed, namely a) the analysis of flood depths, conservation state (from the sensitivity component of FV), and wall orientation (from the exposure component of FV) and b) the overlay of flood depths with the sensitivity component of FV. Complementarily, the risk-matrix analysis identified other buildings as high risk, some of them not coincident with the individualised analysis of risk components (hazard and vulnerability).

It is the ability to characterise the potential impacts of hazardous processes that allows us to prepare and promote the necessary formal and informal changes, which, ultimately, will contribute to risk reduction [23]. In that sense, we expect the generated knowledge to be applicable in different fields of flood risk management.

If the individual first-order results are used, resources' assignment will prove to be more efficient in addressing the particular risk factors identified as more relevant in each building. Combined with the risk-matrix classification, the results are capable of informing municipal decision-makers and planners regarding the intervention priorities of urban rehabilitation projects. Civil protection agents will be capable of planning efficient and safe evacuation routes, in case of flash flooding. The business sector will be able to prepare for recurrent flooding with minor consequences, defined by some as nuisance flooding. Finally, medium- to long-term strategies of spatial planning and design can be drawn from the risk-matrix results: the eventual relocation of buildings classified as high risk; the retrofitting of their physical characteristics in order to reduce their physical vulnerability while maintaining their functionality, if possible, even during flood events, in the areas of low risk. Intermediate contexts of flood risk require, in terms of spatial and design planning, pondering the range of possible interventions, specifically those that better combine urbanity and safety in distinct degrees of flood adaptation [24].

When coupled with social vulnerability data, the provided risk assessment will significantly contribute to increasing the resilience of the built environment. The historical centres of cities represent places of high sensitivity of their exposed elements. In addition to their heritage and cultural value, the mandated authorities need to consider the functions that these buildings provide, both for the resident population and the transient population.

The analysis we have presented, although focused on flood risk, can be replicated with some adaptation, concerning other hazard processes, not exclusively of hydro-geomorphologic origin. Moreover, provided that the evaluated buildings can be grouped typologically, we believe that this framework can be easily applied in larger-scale risk assessments, keeping a very reasonable balance between accuracy and applicability. In that case, blocks of buildings or even entire neighbourhoods can be used as the basic assessment units. An interesting example of using neighbourhoods as the basic assessment unit to evaluate fire risk can be found in [25].

**Author Contributions:** Conceptualisation, T.M.F. and P.P.S.; methodology, T.M.F. and P.P.S.; hazard analysis, P.P.S.; vulnerability analysis, T.M.F.; investigation, T.M.F. and P.P.S.; writing—original draft preparation, T.M.F. and P.P.S.; writing—review and editing, T.M.F. and P.P.S.; funding acquisition, T.M.F. and P.P.S. All authors have read and agreed to the published version of the manuscript.

**Funding:** Tiago M. Ferreira is funded by the Portuguese Foundation for Science and Technology (FCT) through the postdoctoral grant SFRH/BPD/122598/2016 and Pedro P. Santos is funded through the project with the reference CEEIND/00268/2017.

**Acknowledgments:** The authors would like to acknowledge the City Council of Guimarães, particularly the Architect Filipe Fontes, for his generous support and contribution to the development of this work.

**Conflicts of Interest:** The authors declare no conflict of interest.

## References

1. UNISDR. *2009 UNISDR Terminology on Disaster Risk Reduction*; International Strategy for Disaster Reduction: Geneva, Switzerland, 2009; ISBN 978-600-6937-11-3.
2. Rana, I.A.; Routray, J.K. Integrated methodology for flood risk assessment and application in urban communities of Pakistan. *Nat. Hazards* **2018**, *91*, 239–266. [CrossRef]
3. Birkmann, J.; Cardona, O.D.; Carreño, M.L.; Barbat, A.H.; Pelling, M.; Schneiderbauer, S.; Kienberger, S.; Keiler, M.; Alexander, D.; Zeil, P.; et al. Framing vulnerability, risk and societal responses: The MOVE framework. *Nat. Hazards* **2013**, *67*, 193–211. [CrossRef]
4. Ortiz, R.; Ortiz, P.; Martín, J.M.; Vázquez, M.A. A new approach to the assessment of flooding and dampness hazards in cultural heritage, applied to the historic centre of Seville (Spain). *Sci. Total Environ.* **2016**, *551–552*, 546–555. [CrossRef] [PubMed]
5. Wang, J.-J. Flood risk maps to cultural heritage: Measures and process. *J. Cult. Herit.* **2015**, *16*, 210–220. [CrossRef]
6. Miranda, F.N.; Ferreira, T.M. A simplified approach for flood vulnerability assessment of historic sites. *Nat. Hazards* **2019**, *96*, 713–730. [CrossRef]
7. Marzeion, B.; Levermann, A. Loss of cultural world heritage and currently inhabited places to sea-level rise. *Environ. Res. Lett.* **2014**, *9*, 34001. [CrossRef]
8. Granda, S.; Ferreira, T.M. Assessing Vulnerability and Fire Risk in Old Urban Areas: Application to the Historical Centre of Guimarães. *Fire Technol.* **2019**, *55*, 105–127. [CrossRef]
9. Ramísio, P.; Duarte, A.; Vieira, J. Integrated flood management in urban environment: A case study. In Proceedings of the XIV World Water Congress "Bridging Science and Policy", Porto-da-Galinhas, Brazil, 25–29 September 2011.
10. Zêzere, J.L.; Pereira, S.; Tavares, A.O.; Bateira, C.; Trigo, R.M.; Quaresma, I.; Santos, P.P.; Santos, M.; Verde, J. DISASTER: A GIS database on hydro-geomorphologic disasters in Portugal. *Nat. Hazards* **2014**, *72*, 503–532. [CrossRef]

11. Ramísio, P.; Duarte, A.; Vieira, J. *Estudo Hidrológico e Modelação Hidrodinâmica: Estudo hidrológico da bacia hidrográfica da ribeira de Couros*; University of Minho: Guimarães, Portuga, 2011.
12. Ramísio, P.; Duarte, A.; Vieira, J. *Estudo Hidrológico e Modelação Hidrodinâmica: Modelação hidrodinâmica da ribeira de Couros—Situação de referência*; University of Minho: Guimarães, Portugal, 2012.
13. USACE HEC-RAS. River Analysis System. In *Hydraulic Reference Manual*; US Army Corps of Engineers, Hydrologic Engineering Center: Davis, CA, USA, 2016; p. 547. Available online: https://www.hec.usace.army.mil/software/hec-ras/documentation/HEC-RAS%205.0%20Reference%20Manual.pdf (accessed on 8 June 2020).
14. Vicente, R.; Parodi, S.; Lagomarsino, S.; Varum, H.; Silva, J.A.R.M. Seismic vulnerability and risk assessment: Case study of the historic city centre of Coimbra, Portugal. *Bull. Earthq. Eng.* **2011**, *9*, 1067–1096. [CrossRef]
15. Clausen, L. Potential Dam Failure: Estimation of Consequences, and Implications for Planning. Unpublished Master's Thesis, School of Geography and Planning at Middlesex Polytechnic Collaborating with Binnie and Partners, Redhill, UK, 1989.
16. Kelman, I.; Spence, R. An overview of flood actions on buildings. *Eng. Geol.* **2004**, *73*, 297–309. [CrossRef]
17. QGIS Development Team. *QGIS Geographic Information System*; Open Source Geospatial Foundation: Beaverton, OR, USA, 2017.
18. Stephenson, V.; D'Ayala, D. A new approach to flood vulnerability assessment for historic buildings in England. *Nat. Hazards Earth Syst. Sci.* **2014**, *14*, 1035–1048. [CrossRef]
19. Mebarki, A.; Valencia, N.; Salagnac, J.L.; Barroca, B. Flood hazards and masonry constructions: A probabilistic framework for damage, risk and resilience at urban scale. *Nat. Hazards Earth Syst. Sci.* **2012**, *12*, 1799–1809. [CrossRef]
20. McBean, E.A.; Gorrie, J.; Fortin, M.; Ding, J.; Moulton, R. Flood Depth—Damage Curves By Interview Survey. *J. Water Resour. Plan. Manag.* **1988**, *114*, 613–634. [CrossRef]
21. Pistrika, A.; Tsakiris, G.; Nalbantis, I. Flood Depth-Damage Functions for Built Environment. *Environ. Process.* **2014**, *1*, 553–572. [CrossRef]
22. Velasco, M.; Cabello, À.; Russo, B. Flood damage assessment in urban areas. Application to the Raval district of Barcelona using synthetic depth damage curves. *Urban Water J.* **2016**, *13*, 426–440. [CrossRef]
23. Birkmann, J.; Buckle, P.; Jaeger, J.; Pelling, M.; Setiadi, N.; Garschagen, M.; Fernando, N.; Kropp, J. Extreme events and disasters: A window of opportunity for change? Analysis of organisational, institutional and political changes, formal and informal responses after mega-disasters. *Nat. Hazards* **2010**, *55*, 637–655. [CrossRef]
24. Hobeica, L.; Santos, P. Design with floods: From defence against a 'Natural' threat to adaptation to a human-natural process. *Int. J. Saf. Secur. Eng.* **2016**, *6*, 616–626. [CrossRef]
25. Granda, S.; Ferreira, T.M. Large-scale Vulnerability and Fire Risk Assessment of the Historic Centre of Quito, Ecuador. *Int. J. Archit. Herit.* **2019**, 1–15. [CrossRef]

water

MDPI

Article

# Indirect Impact Assessment of Pluvial Flooding in Urban Areas Using a Graph-Based Approach: The Mexico City Case Study

**Marcello Arosio [1,*], Mario L. V. Martina [1], Enrico Creaco [2] and Rui Figueiredo [3]**

1 Department of Sciences, Technologies and Society, Scuola Universitaria Superiore IUSS Pavia, 27100 Pavia, Italy; mario.martina@iusspavia.it
2 Department of Civil Engineering and Architecture, Università degli Studi di Pavia, 27100 Pavia, Italy; creaco@unipv.it
3 CONSTRUCT-LESE Institute of R&D in Structures and Construction—Laboratory of Earthquake and Structural Engineering, Faculty of Engineering, University of Porto, 4200-465 Porto, Portugal; ruifig@fe.up.pt
* Correspondence: marcello.arosio@iusspavia.it

Received: 30 April 2020; Accepted: 16 June 2020; Published: 19 June 2020

**Abstract:** This paper presents the application of a graph-based methodology for the assessment of flood impacts in an urban context. In this methodology, exposed elements are organized as nodes on a graph, which is used to propagate impacts from directly affected nodes to other nodes across graph links. Compared to traditional approaches, the main advantage of the adopted methodology lies in the possibility of identifying and understanding indirect impacts and cascading effects. The application case concerns floods numerically reconstructed in Mexico City in response to rainfall events of increasing return periods. The hazard reconstruction was carried out by using a simplified hydrological/hydraulic model of the urban drainage system, implemented in EPASWMM, the Storm Water Management Model developed by the United States Environmental Protection Agency. The paper shows how the impacts are propagated along different orders of the impact chain for each return period and compares the risk curves between direct and indirect impact. It also highlights the extent to which the reduction in demand of services from consumers and the loss of services from suppliers are respectively contributing to the final indirect impacts. Finally, it illustrates how different impact mitigation measures can be formulated based on systemic information provided by the analysis of graph properties and taking into account indirect impacts.

**Keywords:** pluvial flood; indirect impacts; risk assessment; graph analysis; flood mitigation; Mexico City

## 1. Introduction

Floods have become one of the most dangerous and costly natural hazards in recent decades. Economic damage reported worldwide from 2000 to 2006 add up to more than 422 billion dollars, accompanied by more than 290,000 fatalities and over 1.5 billion affected people [1]. The most dangerous flood events usually take place in urban areas, where the highest number of inhabitants and the most valuable exposed assets are located today, due to current urbanization trends [2]. Another factor contributing to the negative effects of floods is climate change, which appears to be concentrating the total yearly rainfall volume in increasingly sporadic and intense rain events [3]. As a result, rainwater discharges have been growing significantly in urban catchments [4], causing the occurrence of increasingly frequent flood event [5–7]. Due to the rapidity of the governing processes, mainly related to short concentration times, flood events in urban areas typically occur with little to no early warning, thus being commonly referred to as flash floods.

The implementation of effective strategies to manage urban flood risk requires support from risk assessment studies quantifying the impacts of hazardous events on the built environment, economy, and society [8]. The research community concerned with disaster risk reduction (DRR), particularly in the fields of physical and environmental science, has generally agreed on a common approach for the quantification of risk as a function of hazard, exposure, and vulnerability [9]. Hazard defines the potentially damaging events and their probabilities of occurrence, exposure represents the population or assets located in hazard zones that are therefore subject to potential losses, and vulnerability links the intensity of a hazard to potential losses at exposed elements.

When the pluvial flood hazard component of risk is represented using numerical modelling, it is generally assumed to include the ponding prior to the ingress of direct runoff into the underground drainage system (e.g., [10–12]). In terms of floodwater, other contributions come from surcharged sewers and/or from urban minor watercourses, the flow capacity of which has been exceeded as a result of heavy rainfall [13,14]. In most applications, flood hazard (flood extension, water levels, and velocities) is assessed by making use of suitable software (e.g., [15–17]) aiming to reproduce the response of the urban drainage system to intense rain events of prefixed return period (e.g., [18–20]). In addition to rainfall, the presence of close rivers and water courses may be an additional potential driving force for the system [21].

Combining the flood hazard component with exposure and vulnerability data and models enables the computation of flood risk. Methodologies for risk assessment may differ according to the typology of losses considered, which can be divided into direct and indirect, and tangible and intangible [22,23]. Models for the assessment of direct loss and risk are the most widely used, and the ones which have received the most scientific attention in recent decades (e.g., [24,25]). The situation is different regarding indirect losses. In fact, the application of models that take into indirect impacts and cascading effects is much less frequent [26], and the selection and implementation of a model type for estimating indirect economic losses for a given application case is not straightforward [27].

The Economic Global Forum Report (2018) highlights the importance to understand the risk due to the new complexity of our society, and to improve the capacity to model and manage this risk, "When a risk cascades through a complex system, the danger is not of incremental damage but of 'runaway collapse'—or, alternatively, a transition to a new, suboptimal status quo that becomes difficult to escape." Pescaroli and Alexander (2016) [28] showed how the complex socio-technological networks of present-day society, especially in strong urbanized areas, increase the impact of local events on broader crises. These potential evolution processes require a system thinking and perspective that considers all the behaviours of a system as a whole in the context of its environment. The systems perspective uses a non-reductionist approach aiming to describe the properties of the entire system itself. Furthermore, any description of the whole system must include an explanation of the relationships between the parts. Arosio et al. (2020) [29] demonstrate the advantages of representing a complex system, such as an urban settlement, by means of a graph, and using the techniques made available by the branch of mathematics called graph theory. In fact, it is possible to establish analogies between certain graph metrics and the risk variables and to use the graph as a tool to propagate the damage into the system. This approach aims to understand certain risk mechanisms, such as how the impacts of a hazard are propagated. Therefore, the disaster risk of the system as a whole can be assessed, including higher-order impacts and cascading effects.

An outstanding example of drainage system that has been put to a severe test by changes in the last decades is Mexico City (MC). In addition to the effects of climate change and urbanization on flood risk, an extremely high subsidence rate of about 30 centimetres per year [30] is threatening the hydraulic functionality of this drainage system. On top of that, the city has been experiencing significant population growth in the recent decades; thus, making flood management an urgent issue. Given this context, this paper first provides a simplified model that explicitly integrates the drainage system and surface runoff for the estimation of pluvial flood hazard, and then estimates direct and indirect impacts in terms of number of affected graph nodes and links. In particular, the aims of this paper are:

- to set up of a simplified hydrological and hydraulic model of the main urban drainage system to estimate the hazard in terms of flood area extension and water levels;
- to use the graph representation of a system as the basis for assessing higher-order impacts and cascading effects for different return periods (T), based on the propagation of impacts along graph links; and
- to illustrate how different impact mitigation measures can be formulated based on systemic information provided the analysis of graph properties, taking into account indirect impacts.

## 2. Methodology

### *2.1. Hazard Modelling*

Hazard identification and assessment are carried out by using the Storm Water Management Model developed by the United States Environmental Protection Agency (EPASWMM) [31], which has been adopted in many works in the scientific literature (e.g., [32–35]). This software can model the response of an urban catchment to rain events, while representing the external urban surfaces as planes featuring pre-assigned values of area, width, and slope. The runoff leaving the planes enters the underground system through the junctions present at channel ends and is then routed inside the channels up to the outlet(s) using De Saint Venant equations [36]. At each junction, the mass conservation equation is applied to balance entering and outgoing flows. In correspondence to intense rain events, backwater effects can take place, causing junctions to get surcharged. As a result, water goes out of the underground system, causing floods in the urban environment. The thorough assessment of flood extension and levels would require 2D flood propagation models to be used. However, if an urban territory is inside a basin, the propagation effects can be neglected, and flood extension and levels can be simply assessed by distributing flood volumes over territory, starting from the lowest ground elevations. This can be accomplished by using simple Geographic Information System (GIS) procedures [37].

In EPASWMM, the structure of the urban drainage networks needs to be defined in terms of channel paths, lengths, shapes, and sizes. Then, each junction, which can be a node in common between two adjacent channels, the starting node of a channel, or the final outlet of the system, can be connected to an external catchment. Catchments must be associated with a rain gauge, in which hyetographs can be implemented as time series to represent synthetic rain events of different return periods.

### *2.2. Graph Construction*

This work demonstrates an approach to assess impacts beyond direct tangible damage, integrating the hazard analysis with a graph representation of a complex system. While the traditional reductionist risk assessment framework considers each exposed element individually, a graph representation is able to account for interactions between exposed elements at risk, which is key to a more comprehensive understanding of risk.

Graphs can be used to represent physical elements in the Euclidean space (e.g., electric power grids and highways) or entities defined in an intangible space (e.g., collaborations between individuals). Formally, a graph G consists of a finite set of elements V(G) called vertices (or nodes), and a set E(G) of pairs of elements of V(G) called edges (or links) [38]. The mathematical properties of a graph can be studied using graph theory [39,40]. Such properties, e.g., degree, hub and authority, are useful metrics for analysing the graph structure (i.e., network topology, and arrangement of a network) and in the present context may be used to characterize a collection of exposed elements from a systemic viewpoint (e.g., [41]).

The complete procedure to construct a graph representing a system of exposed assets is presented in Arosio et al. (2020) [29] and its two main steps are summarized here. First, the conceptual network is defined by means of two types of network objects, nodes (vertices) and links (edges), and by specifying their characteristics. Nodes can theoretically represent all the entities to be analysed: physical elements like single buildings, fire stations, and electric towers; supply of services such as those provided by schools, hospitals, and fire stations; or even beneficiaries such as population in general, or students, or elderly people more specifically. Links can be of different types according to the nature of the connection: physical, geographical, cyber, or logical [42]. Second, once the network has been conceptually defined, it is necessary to construct the actual connections between all elements. In fact, the conceptual network determines only the existence of connections between categories of elements, it does not define which specific node of one typology is linked to a specific node of another typology. Therefore, it is necessary to define rules that establish connections between single nodes. These connections can be represented either by a list of pairs of nodes or by an adjacency matrix.

### 2.3. Impact Assessment

Arosio et al. (2020) [29] propose analysing the properties of both the entire graph and single nodes. The global properties show how the whole system is vulnerable to an external perturbation, while the properties of single nodes underline which parts of the system are more critical for the entire system. The analysis of properties provides valuable information on the system. However, in order to assess risk and realize a proper DRR strategy, this information needs to be integrated and overlapped with hazard information (e.g., intensity, extension, and probability of occurrence). In the present work, the impact is computed by integrating the hazard for different return periods with the exposed network. The total impact considers both direct impact ($D$) and indirect impact ($I$).

For the direct impact, recent literature considers the overall amount of physical damage to assets ($D$) equal to the cost to repair and/or replace them, i.e., the sum of the cost to restore the damage ($C_n$) of all the nodes $N$ that are hit by the hazard [43]. Under this condition, the direct physical impact is a function of the number of hit nodes ($N$), the hazard intensity ($H_n$) at each node $n$, and its physical vulnerability ($V_n$):

$$D = \sum_{n=1}^{N} C_n(H_n, V_n) \tag{1}$$

The indirect impact ($I$) considers the overall amount of tangible losses generated by cascading effects due to the direct damage ($D$), i.e., the sum of the economic loss ($L_S$) of all the services $S$ that are interrupted. In this specific definition, the indirect impact is a function of the number of services lost ($S$), the typologies of services ($T_s$), and economic values ($V_s$) associated to each typology value.

$$I = \sum_{s=1}^{S} L_s(T_s, V_s) \tag{2}$$

## 3. Application: Mexico City

Mexico City is one of the most hazard-prone cities in the world, due to the frequent occurrence of floods, landslides, subsidence, volcanism, and earthquakes. The Mexico City Metropolitan Area (MCMA), situated in a high mountain valley (approximately 2200 m a.s.l.), is one of the largest urban agglomerations in the world. Located in a closed basin of 9600 km$^2$, MCMA spreads over a surface of 4250 km$^2$. The MCMA has an estimated metropolitan population of about 21.2 million people, equal to 18% of the country's population, and generates 35% of Mexico's gross domestic product [44]. This application study focusses on Mexico City (also called the Federal District—MCFD), where approximately 8.8 million people live. The choice of MCFD as a pilot case allows showing the importance of interdependencies in assessing the total impact in a complex urban environment. Tellman et al. (2018) [45] show how the risk in Mexico City's history has become interconnected and reinforced. In fact, as cities expand spatially and become more interconnected, risk becomes endogenous. Urbanization, driven by population growth, increases the demand for water and land in many parts of the world [46], which is well represented by the choice of Mexico City as case study. Furthermore, the impact estimated in the present work is expected to worsen in the future, due to the soil subsidence phenomenon in progress, which is expected to deteriorate the hydraulic capacity of the drainage system.

### 3.1. Description of the Application

#### 3.1.1. Hazard Simulation

Rainfall patterns associated with different return periods were obtained through the uniform depth duration frequency curve (DDF) for the entire MCFD [47]. In particular, Chicago hyetographs with a duration of 6 h [48] and with peak located at 2.1 h were constructed starting from the DDF curve [49]. Figure 1 presents the DDF curves computed for different return periods, which were used in the analysis.

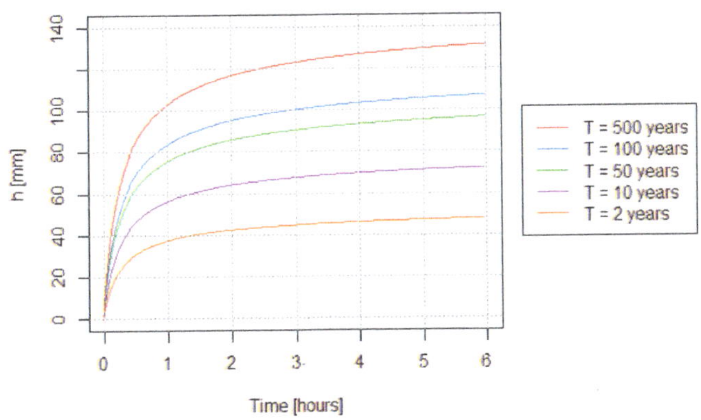


**Figure 1.** Depth duration frequency (DDF) curves for Mexico City.

The EPASWMM model of the main urban drainage system, serving a total area of 1586 km$^2$, was constructed starting from data made available by the Engineering Institute of the National Autonomus University of (UNAM) México D.F. The model has 109 junctions (one for each catchment), and 98 underground channels with Strickler roughness set at 52 m$^{1/3}$/s, the typical value for masonry walls [48] and consider the condition of possible failure of the pumping system (see Figure 2 for the layout of the channels).

The model was used to simulate hydrology and hydraulics in Mexico City, i.e., runoff formation from external catchments and flow-routing in underground channels, respectively, as a result of the rainfall patterns described above. Water volumes going out from each surcharged node of the model were lain on its associated urban catchment, starting from the lowest areas close to the node itself. Incidentally, the neglection of flood propagation over the territory can be considered an acceptable assumption, due to the basin-shaped orography.

By applying the mass conservation equation to balance entering and outgoing flows at each junction, EPASWMM estimates the hydrograph of water that comes out when the piezometric curve exceeds the surface level. Furthermore, in a GIS environment, the volume-depth and depth-area functions for each basin were computed using a five-meter resolution digital terrain model (DTM). Knowing the volume from the outgoing hydrographs and using the functions mentioned above, the flood extension was evaluated independently for each basin at different return periods.

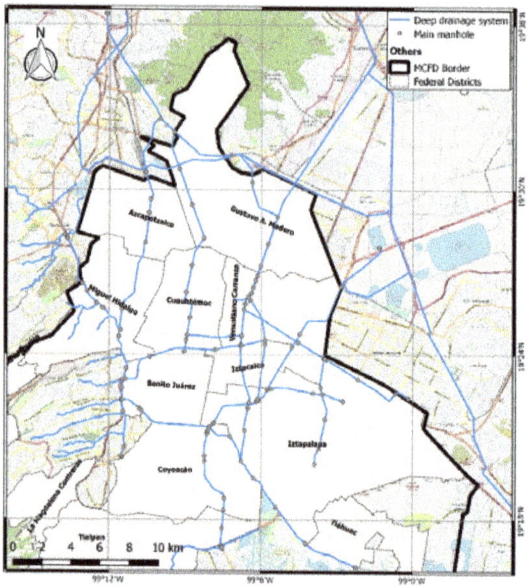


**Figure 2.** Spatial representation of the modelled main drainage system.

### 3.1.2. Network Conceptualization and Graph Construction

The six following critical typologies were selected to for this illustrative pilot study: fire stations, fuel stations, hospitals, schools, crossroads, and blocks [29]. These six typologies do not intend to cover all possible typologies of elements in the city, but rather to illustrate the methodology by representing different types of both tangible and intangible services. The simulation uses blocks instead of population, as this enables a reduction in computational demand by lowering the number of nodes from 8 million to a few tens of thousands. Furthermore, the analysis considers a limited number of central nodes of the city's transportation network (i.e., crossroads) as providers of the transportation service to other elements. This approach does not aim to be representative of the complete behaviour of the road network system (e.g., [50]), but it does allow considering the transportation network in the analysis in a simplified manner. All the typologies, numbers of elements and the connections between them are presented in Table 1. The data utilized to build the network were provided by the Engineer Institute UNAM México D.F.

**Table 1.** List of nodes adopted in the network conceptualization [29].

| Network Conceptualization | Type of Nodes | Number of Elements | Service Provided | Destination of the Service |
|---|---|---|---|---|
| 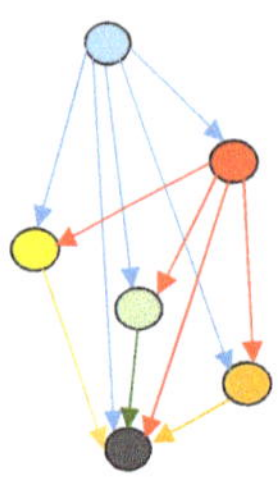 | Crossroads | 17 | Transportation | Fire Stations, Schools, Hospitals, Fuel Stations, and Blocks |
| | Fire Stations | 11 | Recovery | Schools, Hospitals, Fuel Stations, and Blocks |
| | Fuel Stations | 103 | Power | Blocks |
| | Hospitals | 39 | Healthcare | Blocks |
| | Schools | 130 | Education | Blocks |
| | Blocks | 64,282 | - | - |

The mathematical graph, built from the list of nodes and links considered, was obtained using the open source igraph package for network analysis of R environment (http://igraph.org/r/), while the full library of functions adopted are made available by Nepusz and Csard (2018) [51].

#### 3.1.3. Impact Analysis

For the estimation of the direct impact of the nodes, simplified binary vulnerability functions were adopted, consisting of zero damage for the nodes outside of the flooded area, and full damage and loss of functionality for the nodes inside the flooded area. This means that directly impacted nodes are assumed to lose their capacity to provide services, and also their demand for such services. Admittedly, this simple assumption does not realistically represent the estimation of direct damage. However, it is considered acceptable in light of the main focus of this work, which is to investigate and demonstrate the potential of a graph in representing and exploring indirect impacts.

The analysis of the indirect impact refers number of services, as opposed to number of nodes in the direct impact analysis. In fact, indirect impacts are represented here by the difference between the number of the services, i.e., links, existing before the hazardous event and the number of services at the final state, after the cascading effects $\left(S_{initial} - S_{final}\right)$. This analysis could be enhanced with estimates of economic losses corresponding to disrupted services (e.g., related to the fuel sector [52] or the education sector [53]) and/or with economic models to estimate overall economic impacts at a regional scale (e.g., [26]), although these are not carried out here as they are outside the scope of the present article.

The nodes directly impacted by the flood generate a cascading effect into the graph, whereby the propagation of the impact follows the nodes that have lost at least one service from their providers. This propagation process is modelled by creating a new graph ($G_1$) where nodes directly affected by the hazard (i.e., flooded nodes) are removed from the original graph ($G_0$). After that, the degree-in, representative of the number of received services, of each node in $G_1$ is compared with the corresponding degree-in in $G_0$. By deleting all nodes experiencing a degree-in reduction from $G_0$ to $G_1$, graph $G_2$ is obtained. This process is iterated until there are only unaffected nodes in the graph to produce the final graph $G_{final}$, which corresponds, to the final state of the affected system.

In this application, two different indirect impacts (i.e., here measured in term of services lost) are investigated: loss of service providers ($S_P$, i.e., suppliers) and loss of service receivers ($S_R$ i.e., consumers). In fact, after a hazardous event, nodes that provide services can experience a loss of demand ($L_D$) from their receivers ($R$) that are affected by the event. Conversely, the receivers can observe a loss of services ($L_S$) from their providers ($P$) after the hazardous event. The loss of provider services ($S_P$) and of receiver services ($S_R$) can be computed as

$$S_P = \sum_{i=1}^{M} \sum_{j=1}^{N} \omega_{i,j} \tag{3}$$

and

$$S_R = \sum_{i=1}^{M} \sum_{j=1}^{N} \varphi_{i,j}, \tag{4}$$

where $\varphi$ and $\omega$ are respectively the degree-in and degree-out of the $j$th node that at the $i$th order has been deleted; $N$ is the number of deleted nodes at step $i$ and $M$ is the last order of loss propagation (i.e., 4 orders in this application). The loss of demand is computed by summing the degree-in of the deleted nodes, while the loss of services is computed by summing the degree-out of the deleted nodes.

### 3.2. Flood Hazard and Impact Results

This section shows the results obtained from the hydrological and hydraulic simulations, followed by the estimation of direct and indirect impacts. The impacts are previously presented in terms of cascading evolution of affected nodes and later in term of lost connections. It is also presented for each typology of service whether the impact is due to the loss of service provider or of receiver demand and illustrates how different flood mitigation measures can be formulated based on systemic information provided the analysis of graph properties.

The hazard analysis provided the pluvial flood areas associated with the precipitation at each specific return period. As expected, Figure 3 shows that the total flood area increases with growing return period: from 20 km$^2$ at the 2-year return period up to 55 km$^2$ at the 500-year return period.

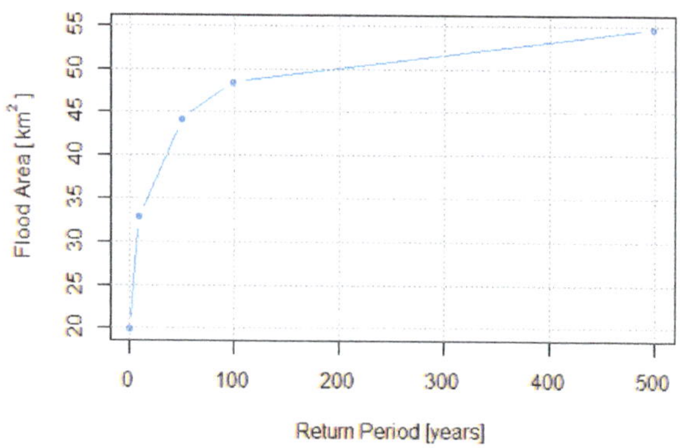


**Figure 3.** Flooded areas for different return periods.

For illustration, Figure 4 presents the areas associated with a flood with the 2- and 500-year return periods, respectively the minimum and maximum considered in this study. In both cases, most water depth values are within realistic ranges (e.g., up to 1.20 m for the 500-year return period flood), with only a limited number of outliers located mostly at the borders of the considered basins. While a more comprehensive flood hazard assessment would be possible by using 2D flood propagation models, these results are considered suitable for the specific scope of this work, which focuses on demonstrating a graph-based approach for higher-order impact assessment and risk mitigation.

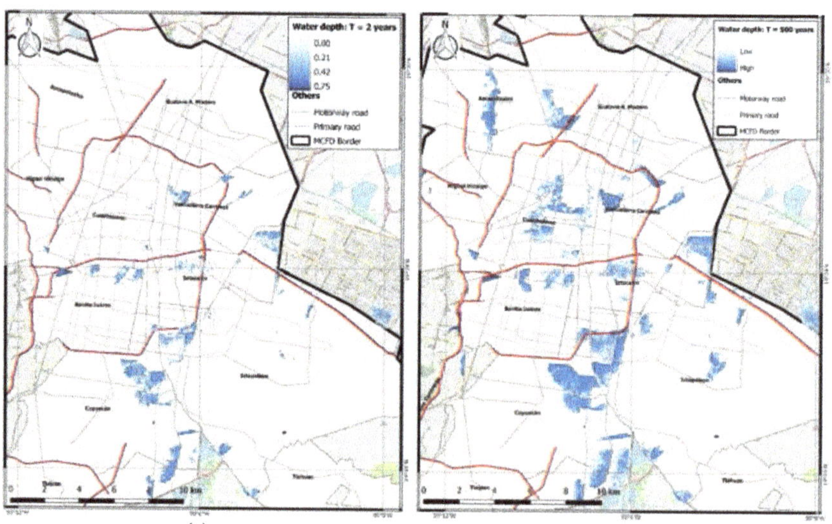

(a) (b)

**Figure 4.** (**a**) Flood map of 2-year return period; (**b**) flood map of the 500-year return period.

Traditional approaches estimate the direct impacts considering only the exposed elements hit by the flood. Instead, the proposed approach can simulate the evolution of cascading effects after some elements are directly hit. Figure 5 shows the evolution of cascading effects of affected nodes at different orders. At each order, the graph was obtained by removing the nodes that were affected in the previous order. Order 0 is the original graph, order 1 is reduced by the nodes that were affected directly by the flood, and from orders 2 to 4 the graphs were obtained by removing indirectly affected nodes, as explained previously. The number of nodes impacted by the event increases when moving from order 1, which represents the directly affected nodes, to order 3, which is associated with the maximum number of nodes affected by the event. The cascading effects stop at order 3 because there are no new nodes affected by parent nodes after that point. Figure 5 shows less than 5000 nodes directly affected for each return period, specifically 3% of the nodes (1818 nodes) and 8% of the nodes (4904 nodes) are flooded for the return periods of 2 years and 500 years, respectively. Instead, the total number of nodes affected at the end of the cascading effects ranges from 23% to 43% of the entire graph, at the 2- and 500-year return periods, respectively.

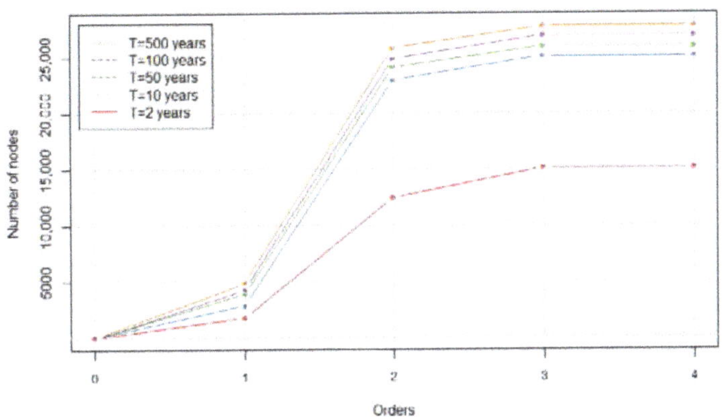


**Figure 5.** Number of nodes affected at each step for different return periods.

Comparing the shape of the curves in Figure 5, a particular behaviour is remarked between the 2- and 10-year return periods when moving from order 1 to 2. In fact, the difference in number of affected nodes between the two return periods is much higher at order 2. The increment of nodes affected at each order, after the direct impact (above order 1), is not proportional with the increment of flooded extension associated with different return period. This is a peculiar aspect of complex systems, in which large-scale secondary events play a not-negligible role.

An explanation for this can be found in Figure 6, which shows the propagation of the effects not only inside the flooded areas, but also outside and potentially over the entire network. Figure 6 shows in red the nodes that are directly impacted (i.e., inside the flood area), in orange the nodes that are indirectly impacted due to the absence of service from the provider nodes inside the flood area (black stars at order 1), and in yellow the nodes at order 3 that are indirectly impacted due to the absence of service from the provider nodes affected at order 2 (grey stars). Since at order 3 there are no new affected providers, the propagation effects stop and the number of impacted nodes at order 4 do not increase. In comparison with the 2-year return period, there is a larger number of nodes affected at order 2 in the 10-year-return period. This happens because the flood directly impacts a hospital, which is not reached by the flood at the 2-year return period. This hospital is an important central node of the graph, which has the capability of influencing many other nodes due to its role of hub. In fact, this node provides the healthcare service to many other nodes, and as shown by the hub analysis carried out in Arosio et al. (2020) [29], it has the highest hub value of the entire system. Figure 6 shows, for the larger return period, more numerous orange nodes (i.e., nodes indirectly impacted at order 2) in the south-east part of the city. This large area, characterized by blocks with higher authority values [29], gets affected due to the absence of service provided by the hospital mentioned above. According to these results, the blocks with higher authority are the ones that depend on the services from the providers with higher values of hubs. For this reason, the floods that hit such hubs undermine a considerable part of the network as is evident for floods with $T$ above 2 years. The strong correlation between hubs and authority explains these results. However, it is necessary to underline that these outputs also reflect the assumption of the rules of proximity adopted in this model, where the network has no redundancy by construction [29]. The redundancy can change the values of hub and authority of the nodes and can therefore influence the magnitude of cascading impacts that are presented in the next section.

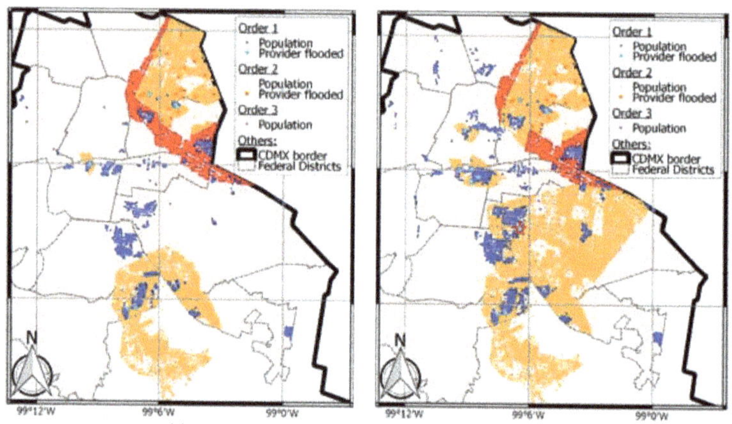


**Figure 6.** Spatial representation of the cascading effects for (**a**) 2- and (**b**) 10-year return periods.

Beyond the evolution of the cascading effects, the comparison between the direct and total impact associated with the flooded areas at different return periods has been investigated and is presented in Figure 7. This figure confirms that while the number of directly impacted nodes is proportional to the flooded area, this is not the case for the total number of impacted nodes, which also considers the indirect impact. As emphasised before, the non-proportional increment is more evident when moving beyond the 10-year return period flood, past which the node with the highest hub value is flooded.

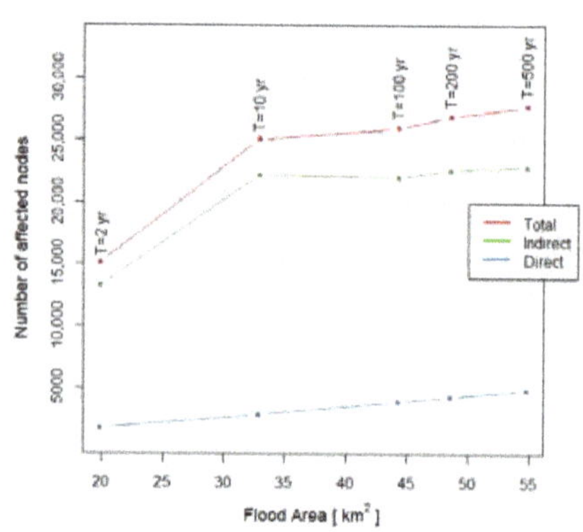


**Figure 7.** Relation between flooded areas at different return periods and number of nodes impacted directly (blue), indirectly (green), and in total (red).

The proposed methodology offers the possibility to investigate not only the nodes but also the connections that are affected during the evolution of the event. This innovative perspective is presented in Figure 8, which shows the probability of observing the loss of services for both direct and indirect impact. In both cases, the curves present the typical shape of risk exceedance probability curves. This figure highlights the much larger numbers of lost services for the indirect impact relative to the direct one, especially for rarer events, where direct impact is only a minor part of the total impact.

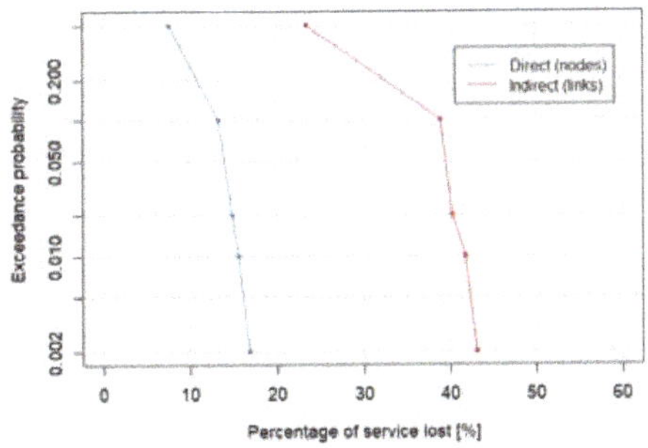


**Figure 8.** Risk curves expressed as percentage of network services lost due to direct and indirect impact.

A more exhaustive explanation of the service lost is provided by the analysis of the causes that generate the lack of services. Figure 9 shows the extent to which the lost services ($\Delta S$) is due to the absence of receivers' demand ($L_D$ on the left) and the absence of offer from providers ($L_O$ on the right). The high values of total demand loss at each return period are mainly driven by the impact on the population; in particular, for the return period of 2 years, the total loss of demand is below 80,000 in comparison with the highest return period featuring a loss of demand above 120,000. The total losses of offers are due to the impacts on the providers and the values are about one third of the total loss of demands. The plot for 10-year return period highlights that the loss of services is mainly due to the absence of service provided by the hospital, which is confirmed by the hub analysis mentioned previously [29]. Conversely, education is the sector across providers that generates more losses, followed by power and healthcare, whereas crossroads and fire stations do not cause loss of demand.

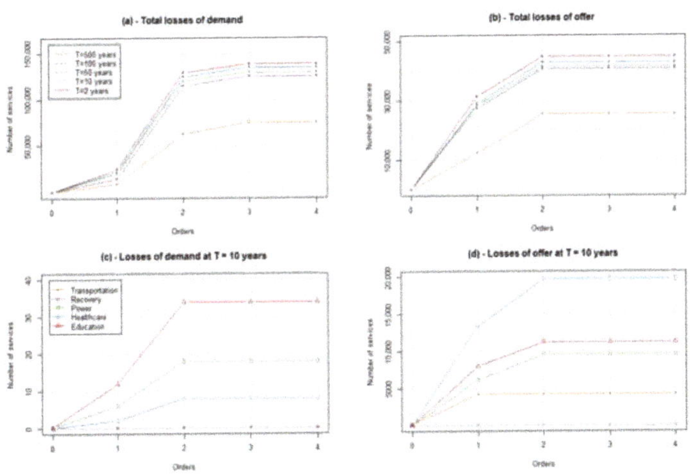


**Figure 9.** Number of services lost at different orders: (**a**,**b**) show the services lost for different return periods; (**c**,**d**) show the services lost at T = 10 years for different typologies.

### 3.3. Mitigation Scenarios

The previous analysis highlights that indirect impacts must not be overlooked when assessing risk. Accordingly, such impacts should be taken into account for the implementation of effective risk mitigation measures, which traditionally tend to be based on the analysis of direct impacts only. The approach used here is able to characterize the system through various graph measures, which can then be used as a starting point for the identification of efficient solutions to mitigate risk, arising not only from direct impacts but also from higher-order effects. To illustrate this, three possible mitigation scenarios (M1, M2, and M3) are proposed. The definition of these hypothetical scenarios was based on information provided by certain graph properties, in particular, hub and authority values, as described next.

Scenario M1—Hazard mitigation: One of the most traditional approaches for mitigating flood risk is to implement measures that reduce the hazard intensity. Therefore, for scenario M1, we consider a hypothetical intervention in the Au De Uarez basin to increase its permeability and thus to reduce the volume of flooding from the drainage system in this specific basin. In the Au De Uarez basin, highlighted in blue in Figure 10, the hospital with the highest value of hub is located. We have considered an intervention that can reduce the waterproof area by up to 60%. As a result, by using equation

$$\Psi = \Psi_P\left(1 - A_{imp}\right) + \Psi_{imp} \times A_{imp} \quad (5)$$

where $\Psi$ is the weighted flow coefficient, $\Psi_P$ and $\Psi_{imp}$ are the flow coefficients respectively for permeable and impermeable area ($A_{imp}$) [54], the reduced flooded volume was calculated. Given the mitigated volume, using the volume–height curve characteristic of the basin, the post-intervention flood heights and extents were obtained. In this new configuration, the hospital is not flooded for any return period.

Scenario M2—Mitigation of the physical vulnerability: Another option to reduce risk is to intervene on the physical vulnerability of the exposed elements. In this alternative, we have chosen three elements that are within the flooded area and have the highest hub values as provided by the graph analysis: a hospital, a school, and a petrol station (highlighted in green in Figure 10). Considering the binary vulnerability functions adopted in this work, reducing the vulnerability means making the three elements waterproof without the loss of functionality. Note that this assumption does not intend to realistically represent the reduction in vulnerability provided by actual waterproofing measures (e.g., [55–57]), but rather to illustrate methodology for this specific scenario.

Scenario M3—Mitigation of the systemic vulnerability: The last scenario is aimed at mitigating the systemic vulnerability associated with the dependency of a large number of nodes on the three service providers identified in the previous scenario (i.e., the nods with the highest hub values). To achieve this, three new providers were added to the network: one hospital, one school, and one petrol station, highlighted in yellow in Figure 10. The choice of location was made considering the area outside the flood extents where nodes present the highest authority values [29].

Figure 11 shows the annual exceedance probability curved (i.e., risk curves), expressed in terms of number of indirectly impacted nodes, for the baseline and the three mitigation scenarios. Regarding scenario M1, the considered mitigation measure is effective for floods with return periods of 10 years and higher because the only benefitted provider is not affected by the 2-year return period flood (i.e., the hospital with hub value equal to 1). Scenario M2 returns the greatest reduction for any return period, which is reasonable given that waterproofing measures were considered for the three elements with the highest hub values. However, in practice, a complete mitigation of physical vulnerability is not attainable, and thus it is likely that flood above a certain return period would affect these elements to some extent. Scenario M3 also leads to an impact reduction for all return periods, although it is not as effective as M2 as implemented in this illustrative analysis. Nevertheless, risk is mitigated to some extent through a redistribution of services provided.

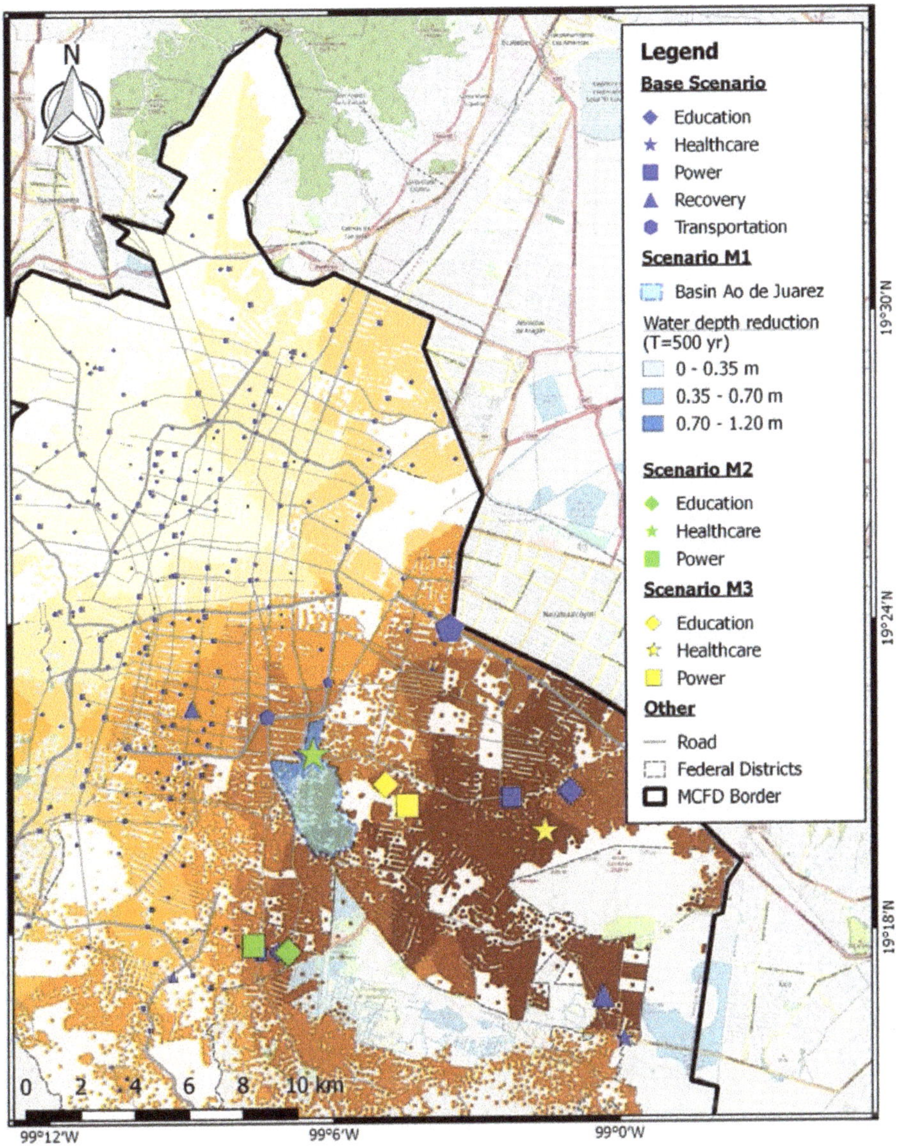


**Figure 10.** Map of the 3 mitigation scenarios. In the base scenario, the orange colour scale used to represent building blocks reflects their authority values, and the symbol size of provider nodes are proportional to their hub values.

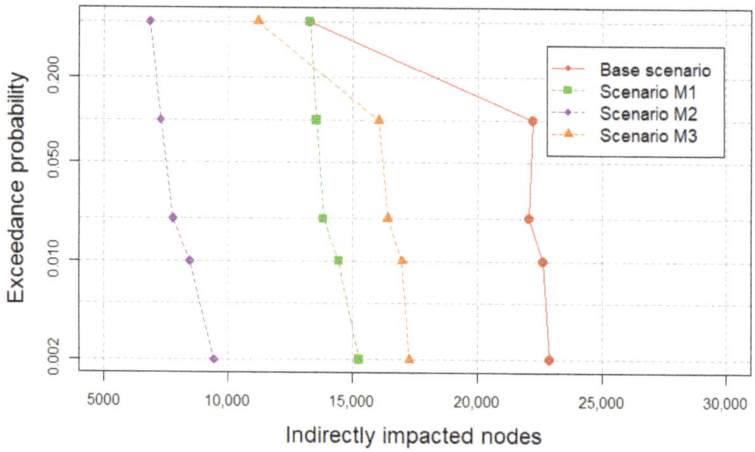


**Figure 11.** Comparison of annual exceedance probability curves for the baseline and three hypothetical mitigation scenarios.

In order analyse the effectiveness of each mitigation scenario, we adopt the most widely-used metric to express risk, which corresponds to the integral of the annual exceedance probability curves shown in Figure 11. This approach may be employed even when losses are not expressed in economic terms [58], as is the case in this study, where our analysis is based on the number of indirectly affected nodes as a proxy for indirect impacts, for the purpose of illustration. For each scenario, we first compute a risk index $R$ as the trapezoidal approximation of the area under the risk curve given by

$$R = \sum_{i=1}^{H-1} (P_{i+1} - P_i) \frac{N_{i+1} - N_i}{2} \tag{6}$$

where $H$ is the number of adopted hazard scenarios, $P_i$ is the annual probability of occurrence of the $i$th hazard scenario (which is assumed to be equal to the inverse of its return period), and $N_i$ corresponds to the number of impacted nodes. We then compute the effectiveness of the different mitigation actions as the reduction of the risk index in relative terms, given by

$$RM[\%] = \frac{R_B - R_j}{R_B} * 100 \tag{7}$$

where $R_B$ is the risk index for the baseline scenario and $R_j$ is the risk index for the $j$th mitigation scenario. Table 2 shows the results for each mitigation scenario considered both individually and in combination, confirming the previously discussed results of Figure 11. In fact, it shows that M2, and any combination that includes it, are the most effective solutions at mitigating the indirect impacts.

**Table 2.** Impact mitigation of the indirect consequences for each scenario.

| Scenario | RM [%] |
|---|---|
| M2 + M1 | 77.93 |
| M2 + M3 | 74.74 |
| M2 | 53.75 |
| M1 + M3 | 45.16 |
| M1 | 24.18 |
| M3 | 20.99 |

## 4. Concluding Remarks

This work focussed on the analysis of the direct and indirect impacts of flood in a complex urban environment that is Mexico City. To this end, we first presented and applied methods for flood hazard assessment at different return periods, due to the spillage of rainwater discharges from surcharged underground drainage channels, under the assumption of pumping systems being out of service. Second, we proposed and applied a graph-based methodology to estimate the total impact of the reconstructed floods and illustrated how graph properties may be used to support the design of mitigation measures.

The results of the work pointed out that, consistently with engineering judgement, flood extensions and levels increase as the return period used for modelling the rainfall grows. Naturally, these are accompanied by the increase of the number of nodes directly affected by the flood. The use of a graph representation of the system then enables the assessment of indirect impacts associated with the flood. The results pointed out that, differently from the direct damage, indirect impacts—and consequently, total impacts—are the result of the number of nodes incrementally affected at each order, which are not proportional to the growth of the flooded area associated with different return periods. This behaviour can be ascribed in this case study to the flooding of an important hub of the graph; thus, generating large cascading effects. Similar situations are expected to occur in other comparable urban environments.

Furthermore, the study showed how the network experiences loss of services at providers ($S_P$) and receivers ($S_R$) for each specific typology of nodes at different orders and return periods. The total losses of service are due to the impacts on the providers and the values are about one third of the total loss of demands, which are mainly driven by the impact on the population. The highest loss of offer is mainly due to the interruption of services provided by the healthcare sector. Conversely, education is the sector that has the highest loss of demand. The impact curves of the lost services in the network showed the extent to which the total impact is larger than the direct impact. In fact, the results of this work confirmed that, especially for extreme rain events, considering only the direct impact leads to a significant underestimation of the total consequences. Furthermore, the last part of the work proved the graph representation of the impacts to be a useful tool for driving decision makers in the choice of the interventions to carry out in the territory, which can include the mitigation of hazard and of physical and systemic vulnerability.

Due to the complexity of the case study and to the unavailability of some data, simplifying assumptions were made along the way, as far as the assessment of hazard, vulnerability, and exposure. In terms of the hazard component, flood extension and levels were assessed by distributing flood volumes over the territory, while neglecting the 2D propagation. Regarding the vulnerability, flood impact was assumed to reduce the services supplied by the directly and indirectly affected nodes immediately to zero through a binary vulnerability function, whereas common sense suggests the loss of functionality to be a growing function of the hazard intensity. Finally, for the exposure component, the risk mitigation measurements adopted are based on the number of nodes and services impacted and not on economic values of the assets and services.

The dynamic evolution from direct to indirect impacts is analysed in this paper by the transformation of the graph at different orders. A more comprehensive analysis should consider the time evolution of the phenomena and not only a progressive order of states. The introduction of the time-element would also allow considering the economic consequences of service interruptions. As an example, regarding the education sector, Sadique et al. [59] investigate the costs of work absenteeism, in terms of paid productivity loss, of care givers who must stay at home during periods of school closure; analogous economic considerations can be done for other sectors. This is outside the scope of the present study, which in this context provides a framework on top of which further economic analyses may be carried out. Future work will be dedicated to incorporating such analyses within the proposed framework, in order to obtain more accurate estimates of the overall impacts of floods in interconnected urban systems.

**Author Contributions:** Conceptualization, M.A., M.L.V.M., E.C. and R.F.; methodology, M.A., M.L.V.M., E.C. and R.F.; software, M.A., M.L.V.M., E.C. and R.F.; formal analysis, M.A., M.L.V.M., E.C. and R.F.; writing—original draft preparation, M.A., M.L.V.M., E.C. and R.F.; writing—review and editing, M.A., M.L.V.M., E.C. and R.F.; visualization, M.A., M.L.V.M., E.C. and R.F.; supervision, M.A., M.L.V.M., E.C. and R.F. All authors have read and agreed to the published version of the manuscript.

**Funding:** This research was partly funded by Fondazione Cariplo under the project "NEWFRAME: NEtWork-based Flood Risk Assessment and Management of Emergencies", and it has been developed within the framework of the project "Dipartimenti di Eccellenza" at IUSS Pavia, funded by the Italian Ministry of Education, University and Research.

**Conflicts of Interest:** The authors declare no conflict of interest. "The funders had no role in the design of the study; in the collection, analyses, or interpretation of data; in the writing of the manuscript, or in the decision to publish the results".

## References

1. Adikari, Y.; Yoshitani, J. *Global Trends in Water-Related Disasters: An Insight for Policymakers—The United Nations World Water Assessment Programme*; The United Nations Educational, Scientific and Cultural Organization: Paris, France, 2009.
2. Jiménez Cisneros, B.E.; Oki, T.; Arnell, N.W.; Benito, G.; Cogley, J.G.; Döll, P.; Jiang, T.; Mwakalila, S.S.; Kundzewicz, Z.; Nishijima, A. Freshwater resources. *Clim. Chang. 2014 Impacts Adapt. Vulnerability Part A Glob. Sect. Asp.* **2015**, 229–270. Available online: https://scholarbank.nus.edu.sg/handle/10635/133175 (accessed on 16 June 2020).
3. Cramer, W.; Yohe, G.W.; Auffhammer, M.; Huggel, C.; Molau, U.; Da Silva Dias, M.A.F.; Solow, A.; Stone, D.A.; Tibig, L.; Leemans, R.; et al. Detection and attribution of observed impacts. *Clim. Chang. 2014 Impacts Adapt. Vulnerability Part A Glob. Sect. Asp.* **2015**, 979–1038. Available online: https://www.zora.uzh.ch/id/eprint/105700/1/2014_HuggelC_WGIIAR5-Chap18_FINAL%20.pdf (accessed on 16 June 2020).
4. Arnell, N.W.; Gosling, S.N. The impacts of climate change on river flood risk at the global scale. *Clim. Chang.* **2016**, *134*, 387–401. [CrossRef]
5. Kreibich, H.; Müller, M.; Thieken, A.H.; Merz, B. Flood precaution of companies and their ability to cope with the flood in August 2002 in Saxony, Germany. *Water Resour. Res.* **2007**, *43*, 1–15. [CrossRef]
6. Zhou, Q.; Mikkelsen, P.S.; Halsnæs, K.; Arnbjerg-nielsen, K. Framework for economic pluvial flood risk assessment considering climate change effects and adaptation benefits. *J. Hydrol.* **2012**, *414–415*, 539–549. [CrossRef]
7. Muis, S.; Güneralp, B.; Jongman, B.; Aerts, J.C.J.H.; Ward, P.J. Science of the Total Environment Flood risk and adaptation strategies under climate change and urban expansion: A probabilistic analysis using global data. *Sci. Total Environ.* **2015**, *538*, 445–457. [CrossRef]
8. Grossi, P.; Kunreuther, H. *Catastrophe Modeling: A New Approach to Managing Risk*; Catastrophe Modeling; Kluwer Academic Publishers: Boston, MA, USA, 2005; Volume 25, ISBN 0-387-23082-3.
9. Balbi, S.; Giupponi, C.; Gain, A.; Mojtahed, V.; Gallina, V.; Torresan, S.; Marcomini, A. The KULTURisk Framework (KR-FWK): A conceptual framework for comprehensive assessment of risk prevention measures—Project deliverable 1.6. *SSRN Electron. J.* **2010**. [CrossRef]
10. DEFRA (Department for Environment Food and Rural Affairs Flood Management Division). *Surface Water Management Plan Technical Guidance*; DEFRA: London, UK, 2010.
11. Parker, D.J.; Priest, S.J.; Mccarthy, S.S. Surface water flood warnings requirements and potential in England and Wales. *Appl. Geogr.* **2011**, *31*, 891–900. [CrossRef]
12. SEPA. *Planning Guidance Strategic Flood Risk Assessment: SEPA Technical Guidance to Support Development Planning*; SEPA: Stirling, Scotland, 2015; pp. 1–21.
13. Pitt, M. *The Pitt Review: Learning Lessons from the 2007 Floods*; Cabinet Office: London, UK, 2008.
14. Ochoa-Rodríguez, S. Urban Pluvial FLood Modelling: Current Theory and Practive. Available online: https://www.raingain.eu/sites/default/files/wp3_review_document_0.pdf (accessed on 16 June 2020).
15. James, W.; Rossman, L.A.; James, W.R.C. *User's Guide to SWMM5*, 13th ed.; CHI: Guelph, ON, Canada, 2010; ISBN 9780980885330.
16. CH2M Flood Modeller Suite. Available online: https://www.floodmodeller.com/ (accessed on 16 June 2020).

17. DHI Mike 2020. Available online: https://www.mikepoweredbydhi.com/mike-2020 (accessed on 16 June 2020).
18. Aronica, G.T.; Candela, A.; Fabio, P.; Santoro, M. Estimation of flood inundation probabilities using global hazard indexes based on hydrodynamic variables. *Phys. Chem. Earth* **2012**, *42–44*, 119–129. [CrossRef]
19. Löwe, R.; Urich, C.; Sto, N.; Mark, O.; Deletic, A.; Arnbjerg-nielsen, K. Assessment of urban pluvial flood risk and efficiency of adaptation options through simulations—A new generation of urban planning tools. *J. Hydrol.* **2017**, *550*, 355–367. [CrossRef]
20. Russo, B.; Sunyer, D.; Velasco, M.; Djordjevic', S. Analysis of extreme fl ooding events through a calibrated 1D/2D coupled model: The case of Barcelona (Spain). *J. Hydroinform.* **2015**, *17*, 473–491. [CrossRef]
21. Galuppini, G.; Quintilliani, C.; Arosio, M.; Barbero, G.; Ghilardi, P.; Manenti, S.; Petaccia, G.; Todeschini, S.; Ciaponi, C.; Martina, M.L.V.; et al. A unified framework for the assessment of multiple source urban flash flood hazard: The case study of Monza, Italy. *Urban Water J.* **2020**, *17*, 35–77. [CrossRef]
22. Penning-rowsell, E.; Johnson, C.; Tunstall, S.; Morris, J.; Chatterton, J.; Green, C.; Koussela, K.; Fernandez-bilbao, A. *The Benefits of Flood and Coastal Risk Management: A Handbook of Assessment Techniques*; Middlesex University Press: London, UK, 2005; ISBN 1 904750 51 6.
23. Merz, B.; Kreibich, H.; Schwarze, R.; Thieken, A. Assessment of economic flood damage. *Nat. Hazards Earth Syst. Sci.* **2010**, *10*, 1697–1724. [CrossRef]
24. Arrighi, C.; Brugioni, M.; Castelli, F.; Franceschini, S.; Mazzanti, B. Urban micro-scale flood risk estimation with parsimonious hydraulic modelling and census data. *Nat. Hazards Earth Syst. Sci.* **2013**, *13*, 1375–1391. [CrossRef]
25. Lüdtke, S.; Schröter, K.; Steinhausen, M.; Weise, L.; Figueiredo, R.; Kreibich, H. A Consistent Approach for Probabilistic Residential Flood Loss Modeling in Europe. *Water Resour. Res.* **2019**, *55*, 10616–10635. [CrossRef]
26. Sieg, T.; Schinko, T.; Vogel, K.; Mechler, R.; Merz, B.; Kreibich, H. Integrated assessment of short-term direct and indirect economic flood impacts including uncertainty quantification. *PLoS ONE* **2019**, *14*. [CrossRef]
27. Okuyama, Y. Economic modeling for disaster impact analysis: Past, present, and future. *Econ. Syst. Res.* **2007**, *19*, 115–124. [CrossRef]
28. Pescaroli, G.; Alexander, D. Critical infrastructure, panarchies and the vulnerability paths of cascading disasters. *Nat. Hazards* **2016**, *82*, 175–192. [CrossRef]
29. Arosio, M.; Martina, M.L.V.; Figueiredo, R. The whole is greater than the sum of its parts: A holistic graph-based assessment approach for natural hazard risk of complex systems. *Nat. Hazards Earth Syst. Sci.* **2020**, *20*, 521–547. [CrossRef]
30. Chaussard, E.; Cabral-cano, E. Magnitude and extent of land subsidence in central Mexico revealed by regional InSAR ALOS time-series survey. *Remote Sens. Environ.* **2014**, *140*, 94–106. [CrossRef]
31. Rossman, L.A. *Storm Water Management Model User's Manual*; National Risk Management Research Laboratory, Office of Research and Development, US Environmental Protection Agency: Cincinnati, OH, USA, 2010.
32. Tsihrintzis, V.; Hamid, R. Runo quality prediction from small urban catchments using SWMM. *Hydrol. Proc.* **1998**, *12*, 311–329. [CrossRef]
33. Guo, J.C.Y.; Urbonas, B. Conversion of Natural Watershed to Kinematic Wave Cascading Plane. *J. Hydrol. Eng.* **2009**, *14*, 839–846. [CrossRef]
34. Burszta-Adamiak, E.; Mrowiec, M. Modelling of Green roofs' hydrologic performance using EPA's SWMM. *Water Sci. Technol.* **2013**. [CrossRef] [PubMed]
35. Rosa, D.J.; Clausen, J.C.; Dietz, M.E. Calibration and verification of SWMM for low impact development. *J. Am. Water Resour. Assoc.* **2015**, *51*, 746–757. [CrossRef]
36. Chaudhry, M.H. *Open-Channel Flow*; Springer: New York, NY, USA, 2008; ISBN 978-0-387-30174-7.
37. Meng, X.; Zhang, M.; Wen, J.; Du, S.; Xu, H.; Wang, L.; Yang, Y. A simple GIS-based model for urban rainstorm inundation simulation. *Sustainability* **2019**, *11*, 2830. [CrossRef]
38. Boccaletti, S.; Latora, V.; Moreno, Y.; Chavez, M.; Hwang, D.U. Complex networks: Structure and dynamics. *Phys. Rep.* **2006**, *424*, 175–308. [CrossRef]
39. Biggs, N.L.; Lloyd, E.K.; Wilson, R.J. *Graph Theory 1736-1936*; Clarendon Press: Oxford, UK, 1976; ISBN 0198539169.
40. Barabasi, A.L. Network Science. Available online: http://networksciencebook.com/ (accessed on 16 June 2020).

41. Newman, M.E.J. *Networks An Introduction*; Oxford University Press: New York, NY, USA, 2010; ISBN 9780199206650.
42. Rinaldi, S.M.; Peerenboom, J.P.; Kelly, T.K. Identifying, understanding, and analyzing critical infrastructure interdependencies. *IEEE Control Syst. Mag.* **2001**, *21*, 11–25.
43. Dottori, F.; Figueiredo, R.; Martina, M.; Molinari, D.; Scorzini, A.R. INSYDE: A synthetic, probabilistic flood damage model based on explicit cost analysis. *Nat. Hazards Earth Syst. Sci.* **2016**, *16*, 2577–2591. [CrossRef]
44. Campillo, G.; Dickson, E.; Leon, C.; Goicoechea, A. *Urban Risk Assessment Mexico City Metropolitan Area, Mexico*; World Bank: New York, NY, USA, 2011.
45. Tellman, B.; Bausch, J.C.; Eakin, H.; Anderies, J.M.; Mazari-Hiriart, M.; Manuel-Navarrete, D.; Redman, C.L. Adaptive pathways and coupled infrastructure: Seven centuries of adaptation to water risk and the production of vulnerability in Mexico City. *Ecol. Soc.* **2018**, *23*, art1. [CrossRef]
46. Odeh, T.; Mohammad, A.H.; Hussein, H.; Ismail, M.; Almomani, T. Over-pumping of groundwater in Irbid governorate, northern Jordan: A conceptual model to analyze the effects of urbanization and agricultural activities on groundwater levels and salinity. *Environ. Earth Sci.* **2019**, *78*, 40. [CrossRef]
47. Amaro, P. *Proposal of an Approach for Estimating Relationships I-d-Tr from 24 Hours Rainfalls*; Universidad Nacional Autónoma de México (UNAM): Mexico City, Mexico, 2005.
48. Artina, S.; Calenda, G.; Calomino, F.; Loggia, G.L.; Modica, C.; Paoletti, A.; Papiri, S.; Rasulo, G.; Veltri, P. *Sistemi di Fognatura. Manuale di Progettazione*; Hoepli: Milano, Italy, 1997.
49. Torres, M.A.; Reinoso, E.; Jaimes, M.A.; De Ingeniería, I.; México, D.F. *Flood Loss Scenarios in Mexico City to Possible Failures of the Main Deep Drainage System*; Universidad Nacional Autónoma de México (UNAM): Mexico City, Mexico, 2012.
50. Pregnolato, M.; Ford, A.; Robson, C.; Glenis, V.; Barr, S.; Dawson, R. Assessing urban strategies for reducing the impacts of extreme weather on infrastructure networks. *R. Soc. Open Sci.* **2016**, *3*, 160023. [CrossRef]
51. Nepusz, T.; Csard, G. *Package "igraph"—Network Analysis and Visualization*. Available online: https://cran.r-project.org/web/packages/igraph/igraph.pdf (accessed on 16 June 2020).
52. Carrera, L.; Standardi, G.; Bosello, F.; Mysiak, J. Assessing direct and indirect economic impacts of a flood event through the integration of spatial and computable general equilibrium modelling. *Environ. Model. Softw.* **2015**, *63*, 109–122. [CrossRef]
53. Borse, R.H.; Behravesh, C.B.; Dumanovsky, T.; Zucker, J.R.; Swerdlow, D.; Edelson, P.; Choe-Castillo, J.; Meltzer, M.I. Closing schools in response to the 2009 pandemic influenza a H1N1 virus in New York City: Economic impact on households. *Clin. Infect. Dis.* **2011**, *52*, 168–172. [CrossRef] [PubMed]
54. Gisonni, C.G. Rasulo Metodi per determinare le massime portate pluviali. In *Sistemi di Fognatura. Manuale di Progettazione*; Hoepli: Milano, Italy, 1997; p. 966, ISBN 9788578110796.
55. Mendes, P.; Lopes, J.G.; De Brito, J.; Feiteira, J. Waterproofing of concrete foundations. *J. Perform. Constr. Facil.* **2014**, *28*, 242–249. [CrossRef]
56. Tong, Z. Research on Waterproof Technology of Construction Engineering. In Proceedings of the International Conference on Chemical, Material and Food Engineering, Kunming, China, 25–26 July 2015; Atlantis Press: Paris, France, 2015; pp. 500–502.
57. Ginda, G. How to help the choice of waterproof insulation to be sustainable? In Proceedings of the 13th International Conference "Modern Building Materials, Structures and Technique", Vilnius, Lithuania, 16–17 May 2019.
58. Figueiredo, R.; Romão, X.; Paupério, E. Flood risk assessment of cultural heritage at large spatial scales: Framework and application to mainland Portugal. *J. Cult. Herit.* **2020**, *43*, 163–174. [CrossRef]
59. Sadique, M.Z.; Adams, E.J.; Edmunds, W.J. Estimating the costs of school closure for mitigating an influenza pandemic. *BMC Public Health* **2008**, *8*. [CrossRef] [PubMed]

MDPI
St. Alban-Anlage 66
4052 Basel
Switzerland
Tel. +41 61 683 77 34
Fax +41 61 302 89 18
www.mdpi.com

*Water* Editorial Office
E-mail: water@mdpi.com
www.mdpi.com/journal/water

www.ingramcontent.com/pod-product-compliance
Lightning Source LLC
Chambersburg PA
CBHW061512210926
51850CB00008B/1897